21st Century Natural Philosophy of Octonion Cosmology, and Predestination, Fate, and Free Will

Stephen Blaha Ph. D.
Blaha Research

Pingree-Hill Publishing

MMXXI

ISBN: 978-1-7372640-6-4

Rev. 00/00/01   September 17, 2021

To Margaret

Some Other Books by Stephen Blaha

All the Megaverse! Starships Exploring the Endless Universes of the Cosmos using the Baryonic Force (Blaha Research, Auburn, NH, 2014)

SuperCivilizations: Civilizations as Superorganisms (McMann-Fisher Publishing, Auburn, NH, 2010)

All the Universe! Faster Than Light Tachyon Quark Starships & Particle Accelerators with the LHC as a Prototype Starship Drive Scientific Edition (Pingree-Hill Publishing, Auburn, NH, 2011).

Unification of God Theory and Unified SuperStandard Model THIRD EDITION (Pingree Hill Publishing, Auburn, NH, 2018).

The Exact QED Calculation of the Fine Structure Constant Implies ALL 4D Universes have the Same Physics/Life Prospects (Pingree Hill Publishing, Auburn, NH, 2019).

Unified SuperStandard Theory and the SuperUniverse Model: The Foundation of Science (Pingree Hill Publishing, Auburn, NH, 2018).

Quaternion Unified SuperStandard Theory (The QUeST) and Megaverse Octonion SuperStandard Theory (MOST) (Pingree Hill Publishing, Auburn, NH, 2020).

Unified SuperStandard Theories for Quaternion Universes & The Octonion Megaverse (Pingree Hill Publishing, Auburn, NH, 2020).

The Essence of Eternity: Quaternion & Octonion SuperStandard Theories (Pingree Hill Publishing, Auburn, NH, 2020).

A Very Conscious Universe (Pingree Hill Publishing, Auburn, NH, 2020).

From Octonion Cosmology to the Unified SuperStandard Theory of Particles (Pingree Hill Publishing, Auburn, NH, 2020).

Beyond Octonion Cosmology (Pingree Hill Publishing, Auburn, NH, 2021).

Available on Amazon.com, bn.com, Amazon.co.uk and other international web sites as well as at better bookstores.

CONTENTS

FIGURES and TABLES

Introduction

Natural Philosophy is based on a current understanding of Physical Reality. Thus it must evolve as our knowledge of Nature changes. This was true in Ancient Greece. It remains true up to the present. We can classify views of Physical Reality into four periods that correspond to major periods in Natural Philosophy:

> Aristotelian
> Newtonian
> Atomic
> Octonion Cosmological

This book describes Natural Philosophy in view of the exciting new developments of Physical Reality in Octonion Cosmology. It begins with a detailed narrative outline of the genesis of the ten spaces (universes) of the Cosmos from a point in a 20 dimension space-time.

It is followed by a detailed view of the Octonion Cosmological Period describing the impact on Epistemology and Metaphysics. Space, time and matter are intertwined in a new manner that differs from the Atomic Period. The structure of the Cosmos differs dramatically from previous theories.

The book also addresses Predestination, Fate, and Free Will from a quantum perspective.

Its third part contains the original book: *21ˢᵗ Century Natural Philosophy*, which describes the Natural Philosophy of the Atomic Period.

The Atomic Period began in the late 19ᵗʰ Century when the atomic nature of matter was finally accepted. It continued up to the 2020 as atoms were disassembled into elementary particles and forces.

In January, 2020 the author introduced a hypercomplex number based theory, QUeST, that had the remarkable feature of yielding the known interactions and elementary particles of The Standard Model and much more. Subsequently Octonion Cosmology emerged. It is described in 2021 in the author's books: *Quantum Space Theory* and *Beyond Octonion Cosmology*.

A point of interest is a "golden" triangle of transformations between mass, energy, and space. Mass can be transformed into energy and vice versa. Mass and/or energy can be transformed into space and vice versa.

PREFATORY NOTE

This author is somewhat surprised to see other Physicists checking his results in previous books, finding them correct, and then proceeding to claim that they are "keeping even" with the author, and entitled to equal credit. As if their lack of independent inspiration and of independent work did not matter.

Traditionally, a physicist takes credit only for his/her independent work—not for simply checking someone else's work.

In recent times this "shoddy" approach to independent work has led to plagiarism on a significant scale as well as intellectual property theft by some physicists (with the occasional help of criminals) to "keep even." *The integrity of Physics is thus impugned and made suspect.*

The perpetrators gain, recognition they do not deserve, that should rightly go to creators. The end result is status given to perpetrators that enables them to undeservedly influence the future development of Physics to its detriment.

PART I: NATURAL PHILOSOPHY OF OCTONION COSMOLOGY

1. Outline of Octonion Cosmology

This chapter outlines the theory of Octonion Cosmology. It is based on the author's books: *Quantum Space Theory With Application to Octonion Cosmology & Possibly To Fermionic Condensed Matter* and *Beyond Octonion Cosmology* as well as previous books by the author. The discussion will be in a narrative form to make it accessible to a wider range of readers.

Subsequent chapters describe new features of Natural Philosophy that emerge from the consideration of Octonion Cosmology.

1.1 Fundamental Required Features

Certain features are necessary for a satisfactory modern theory of fundamental Physics. These features include:

1. Space-times within which dynamic processes can take place.
2. A satisfactory Quantum formalism.
3. Matter.
4. Energy.
5. A space of internal symmetries for SU(3) and so on.

In addition to these necessary requirements are a viable Quantum Field Theory, and a satisfactory Quantum Space Theory[1] of fermion and boson

---

[1] See Blaha (2021g).

quantum fields containing subspaces, and describing the generation and features of these spaces.

1.2 The Primary Space of Octonion Cosmology

Octonion Cosmology begins with the definition of a primary space,[2] which we will call space 0, from which the other spaces evolve. Space 0 has a twenty dimension space-time (nineteen spatial dimensions and one time dimension.) It contains $2^{22} - 20 = 4,194,284$ additional dimensions for internal symmetries such as SU(2)⊗U(1) and SU(3) symmetries.

A space is a universe. Each universe has a space-time with dimensions ranging from zero to eighteen. In addition it has dimensions for irreducible representations of the internal symmetry groups that it supports. Examples of groups are SU(3) – the Strong interactions and SU(2)⊗U(1) – the ElectroWeak interactions. We will use the terms space and universe interchangeably.

We view space 0 as initially having a certain size in both space and time. In that space-time a fermion has a *spinor array* with row and column widths of 1024×1024 and thus 1,048,576 entries. We assume a fermion(s) exists in the 20 space-time dimension space 0.

We now contract the space-time to a 20 dimension point. The fermion spinor array size (rows and columns) remains the same and generates a spinor array. A *spinor space* with 1,048,576 dimensions is thus defined.[3] This space is Cayley-Dickson number 10 in Fig. 1.1. This space is a universe.

---

[2] The space corresponds to Cayley number 2048. See Fig. 1.1 or Blaha (2021c) for details

[3] The transformation of a fermion spinor array to a dimension array is described in detail in the case of fermion annihilation in chapter 6 of Blaha (2021g).

1.3 Space 10

A fermion of the 20 space-time dimension space 0 contains a spinor space (space 10) populated with matter and energy due to its mass. The spinor space, with its elementary particles and energies, is the space[4] of a higher dimension universe. It has an 18 dimension space-time. The choice of 18 dimensions is motivated by the goal of having a Cayley-Dickson spectrum of spaces.

In space 10 a fermion-antifermion annihilation can take place. It can be visualized with the following Feynman-like diagram:[5]

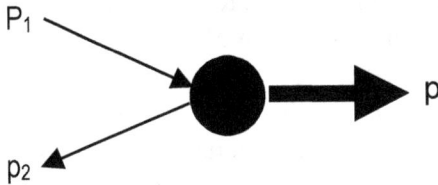

where p_1 and p_2 are the momentum of the fermion and antifermion respectively and p is the momentum of the produced space-containing particle. The produced (scalar) space particle, which exists in space 10,[6] contains a spinor array/space that is the space of another higher dimension universe of the type of space 9 in Fig. 1.1.

Space 9 with its mass and energy is created using the mass and energy from the annihilation in space 10. It has $512 \times 512 = 262,144$ dimensions, of which 16 are space-time dimensions.

---

[4] See Fig. 1.1.
[5] We assume off shell fermions.
[6] Note the produced particle has a mass and momentum in space 10.

This space-time is the first of the series of nine spaces (universes) generated through fermion-antifermion annihilations. They produce a Cayley-Dickson number based spectrum of spaces. See Figs. 1.1 and 1.2 below.

1.4 The Further Generation of Spaces

The spectrum of spaces (universes) is generated through a series of fermion-antifermion annihilations in space after space. The Cayley-Dickson pattern of spaces emerges from the choice of the number of space-time dimensions in each space. We descend from space to space by decreasing the space-time dimension by two at each space.[7] The below Fig. A shows the general pattern of nesting of space instances.[8]

The result of the sequence of "annihilations" is the spectrum of Fig. 1.1 which corresponds to the Cayley-Dickson numbered spaces from 10 through 1 with space-time dimensions from 18 through 0.

In this iterative process there can be any number of subspace instances generated from a space instance by fermion-antifermion annihilations. See Fig. 1.5.

---

[7] An alternate ten space spectrum can be obtained by a diferent set of space-time dimensions. See Fig. 8.2 in Blaha (2021g).
[8] The definition of an instance appears below in the "Point of Clarification."

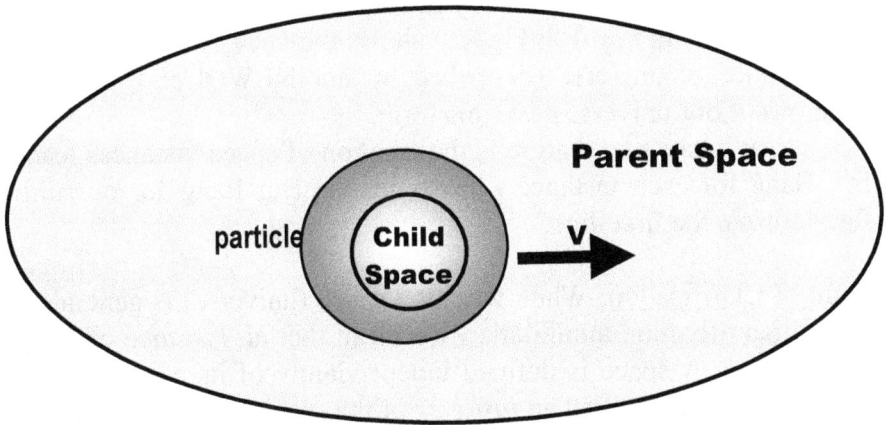

Figure A. The general pattern of a (parent) space containing a particle of velocity **V** that has an internal space instance (child) of lower dimension and lower Cayley-Dickson number. The space is itself within an instance of a higher space. The spectrum's spaces form a nested sequence. Each space contains the next lower space within it. The instances of spaces also form a nested sequence as shown in Fig. 1.2 below.

1.5 Comments

One might expect that generating the sequence of spaces would take an extraordinarily long time. However time, as we know it, only exists in our space 4 universe. Time is not a problem.

The octonion spaces have many dimensions beyond their space-time dimensions. These additional dimensions furnish the fundamental representations of the internal symmetry groups in the space. We show this for space 4 in Fig. 1.3. Fig. 1.4 shows the *form* of the fermion part of the instance of universe (described by our NEWQUeST theory.) The fermions of our universe have this form.

An interesting feature is the creation of space instances leads to a Big Bang for each instance—justifying the Big Bang for our universe for, perhaps, the first time.[9]

Point of Clarification: When we say a space (universe) is generated by a fermion-antifermion annihilation we mean that an *instance* of the space is generated. A space is defined independently of its contents. A space plus its contents is called an *instance* of the space. Annihilations generate instances. The source of the matter and energy of an instance is the mass-energy of the annihilating particles. We sometimes say space to refer to a space instance to avoid cumbersome text.

1.6 The Golden Triangle of Mass, Energy and Space

Examining the Octonion Cosmology spectrum and its origin in fermion-antifermion annihilations, we see a "golden" triangle of transformations between mass, energy, and space. Mass can be transformed into energy and vice versa.

Mass and/or energy can also be transformed into a space and vice versa. The creation of a space (instance) is the result of a mass and energy transformation. A space instance generates mass and energy transformations as it evolves. A space instance also generates instances

---

[99] See Blaha (2021d) for the author's model for universe expansion and the Hubble Constant.

of particles and of subspaces through fermion-antifermion annihilations within it.

Space

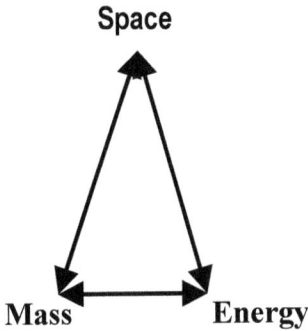

Mass   **Energy**

Figure B. The golden triangle of mass, energy, and space.

A remarkable reflection of the connection between mass and space is the equality of the number of fundamental fermion types in each of the ten spaces and the dimension of the space. In Cayley-Dickson number 4 space, which is that of our universe, we have found 256 fundamental fermion types matching the 16×16 = 256 dimensions of the space.

1.6 Octonion Cosmology Connection to Elementary Particle Theory

The Octonion Cosmology with its ten spaces originated in the author's Unified SuperStandard Theory (UST) that was developed over a number of years, and described in:

Blaha, 2018e, *Unification of God Theory and Unified SuperStandard Model THIRD EDITION*

Blaha, 2020a, *Quaternion Unified SuperStandard Theory (The QUeST) and Megaverse Octonion SuperStandard Theory (MOST)*

Blaha, 2020c, *Unified SuperStandard Theories for Quaternion Universes & The Octonion Megaverse*

Blaha, 2021c, *Beyond Octonion Cosmology*

Blaha, 2021d, *Universes are Particles*

Blaha, 2021e, *Octonion-like dna-based life, Universe expansion is decay, Emerging New Physics*

Blaha, 2021f, *The Science of Creation New Quantum Field Theory of Spaces*

In the Fall of 2019 the author considered the possibility that the UST was based on a deeper theory and proceeded to consider theories based on hypercomplex numbers such as quaternions and octonions. *Thus Octonion Cosmology did not "emerge out of thin air."* It provided a deeper base for the UST.

Remarkably the author found that the hypercomplex theories, QUeST and NEWQUeST, had the same internal symmetry groups, and fundamental fermions, as UST. In addition a hypercomplex theory NEWUTMOST furnished an acceptable theory of the Megaverse (Multiverse). Over the following two years 2020 and 2021 Octonion Cosmology emerged with its ten spaces.

The history of this effort, which has some relevance to Natural Philosophy, appears below. It is extracted from *Quantum Space Theory With Application to Octonion Cosmology Beyond Octonion Cosmology*. All **references below in this section 1.6 are to the contents of that book.**

EXTRACT FROM CHAPTERS 11 & 15 of *Quantum Space Theory*

The [UST and NEWUST] theories originated in the past twenty years from the Standard Model of Particles with $SU(2) \otimes U(1) \otimes SU(3)$, and Two-Tier Quantum and PseudoQuantum Field Theory. Noting the presence of conserved particle numbers, and the presence of at least three fermion generations, we introduced the U(4) Generation Groups and the U(4) Layer Groups together with four layers of four generation fermions. The "Normal" fermions had a corresponding set of four layers of four generations of "Dark" fermions. The result was the Unified SuperStandard Theory (UST) symmetry:

$$\{[SU(2) \otimes U(1) \otimes SU(3)]^2 \otimes U(4)^4\}^4$$

with an additional Strong Interaction U(1) group (analogous to that of ElectroWeak theory) found to be needed. Space-time had four dimensions.

In the Fall, 2019 the author discovered that an octonion-based theory that he constructed and named QUeST had the same internal symmetries as UST with the addition of $U(1)^8$. QUeST's internal symmetry, which could be based on a 16×16 dimension array, was

$$[SU(2) \otimes U(1) \otimes SU(3) \otimes U(1)]^8 \otimes U(4)^{16}$$

The addition of $U(1)^8$ indicates that the Strong Interactions in the theory are broken Strong SU(4).

During 2020 the author developed an octonion space spectrum for both universes and Megaverses, and other spaces. The spectrum was shown to arise from a generation mechanism whereby fermion-antifermion annihilation in a higher space produced an instance of a lower space. A critical part of the derivation of the octonion spectrum was the realization that even space-time dimension spinor arrays are composed of Cayley number rows and columns. Spinor arrays of annihilating fermion-antifermion pairs were shown to generate the dimension arrays of subspace instances.

Analyzing the spinor arrays the author noted that the dimension array could be viewed as composed of 64 dimension blocks, which were further subdivided into 16 dimension subblocks.

The subblock structuring, using the known contents of the Standard Model plus Generation and Layer groups, gives the dimension array structure containing 4×4 subblocks in Figs. 11.1 and 11.2.

Thus there was a *most* satisfactory match between UST and QUeST with the only significant difference being the space-time: four octonion (complex quaternion) coordinates for QUeST and four real space-time coordinates for UST. This difference was resolved in NEWUST and NEWQUeST. See the below Connection groups [discussion].

The form of the square spinor arrays generated by fermion-antifermion annihilation gives 64 dimension blocks and 16 dimension blocks as well as 32 dimension composite blocks that are evidenced in the NEWQUeST (and NEWUTMOST) fermion spectrums and internal symmetry group structure.

Since we see only real dimensions in Reality, we transferred 28 QUeST dimensions from space-time to $U(2)^7$ internal symmetry dimensions. The set of internal symmetries was increased by $U(2)^7$, which we call Connection Groups. Each Connection group specifies interactions between corresponding fermions (e with e, q with q, and so on) in separate layers and between Normal and Dark fermions. The connections between the various blocks of fermions are shown in Fig. 11.3. *We implement the very practical rule that all blocks must be connected by interactions or they would not be of physical interest. A totally isolated block effectively does not exist physically (except possibly for gravitation effects).*

The interactions of the Connection groups must be very weak and/or their gauge bosons must be very massive.

The addition of the Connection Groups and the reduction of space-time dimensions accordingly results in NEWQUeST and NEWUST as summarized below. See appendix 11-A for a discussion of NEWQUeST groups.

Note: the Generation, Layer, and Connection groups are all badly broken. Their vector bosons must be very massive since they have not been detected in experiments.

11.1.1 Internal Symmetries

The groups are ElectroWeak SU(2)⊗U(1), Strong SU(3), Generation Group U(4), Layer Group U(4), and U(2) and U(4) Connection groups obtained by transfer from space-time coordinates (See Blaha 2012c). The SU(3)⊗U(1) symmetry may be a broken SU(4) symmetry in eqs. 11.1 – 11.4. The internal symmetries for the theories are:

UST

$$[SU(2) \otimes U(1) \otimes SU(3)]^8 \otimes U(4)^{16} \qquad (11.1)$$

QUeST

$$[SU(2) \otimes U(1) \otimes SU(3) \otimes U(1)]^8 \otimes U(4)^{16} \qquad (11.2)$$

NEWQUeST

$$[SU(2) \otimes U(1) \otimes SU(3) \otimes U(1)]^8 \otimes U(4)^{16} \otimes U(2)^7 \qquad (11.3)$$

The only change is in internal symmetries: Twenty-eight real dimensions transferred from space-time coordinates to Connection group $U(2)^7$ internal symmetry.

NEWUST

$$[SU(2) \otimes U(1) \otimes SU(3) \otimes U(1)]^8 \otimes U(4)^{16} \otimes U(2)^7 \qquad (11.4)$$

The only change in internal symmetries: Twenty-eight real dimensions added for $U(2)^7$ Connection group internal symmetry.

11.1.2 Space-Time Coordinates
UST

Four real space-time coordinates.

QUeST

Four octonion (complex quaternion) coordinates.

NEWUST

Four real space-time coordinates. No change from UST space-time.

NEWQUeST

Four real space-time coordinates. The six coordinates in the $n = 4$ octonion space (Figs. 1.5 and 1.6) were lowered to four space-time coordinates with two coordinates transferred to Connection groups.

The only change is in space-time coordinates: Fourteen dimensions transferred from QUeST space-time coordinates to Connection group $U(2)^7$ internal symmetry.

11.2 Fundamental Fermion Spectrum

There are 256 fundamental fermions[10] in NEWQUeST and NEWUST. Conceptually their structure can be viewed as an extrapolation of the known three generations of The Standard Model. For reasons given previously and in Appendix 11-A a fourth generation was indicated and a corresponding Dark sector of similar structure was added. In addition, because of the need for Layer groups (Appendix 11-A), the overall structure consisted of four copies of this layer.

---

[10] See Fig. 9.2b for the fermion spectrum determined by a spinor array.

Correspondingly, each layer also has its own set of internal symmetry gauge groups[11] to limit mixing between the layers to Layer group interactions and Connection group interactions.

Fig. 11.4 shows the structure of the NEWQUeST/NEWUST fermions. The blocks are 4 × 4 reflecting the origin of the NEWQUeST/NEWUST space (universe) instance from Megaverse fermion-antifermion annihilation. The spinor analysis of their spinor arrays yields a 16 dimension-based block structure. The 64 dimension fermion layers reflect the 64 dimension structuring of the Megaverse obtained from its creation by fermion-antifermion creation in the Maxiverse.

11.3 Total Dimensions

The total of internal symmetry and space-time dimensions is 256 in all four theories listed above. It is based on the 16 × 16 dimension array of the Cayley number n = 4 space of the Octonion spectrum.

11.4 Pattern of Internal Symmetries

The NEWQUeST dimension array for internal symmetries is subdivided into four layers of 56 dimensions—just as in NEWUST (and UST). Fig. 11.1 displays the layers using SU(4) in place of SU(3)⊗U(1). Each layer has a block of 28 dimensions for Normal and 28 dimensions for Dark sectors. There are also seven U(2) Connection groups[12] plus four real-valued space-time coordinates bringing the NEWQUeST total to 256 dimensions.= 4*56 + 28 + 4 = 256 dimensions. The Connection

---

[11] See Fig. 9.2c for the pattern of symmetry groups.
[12] See Appendix 11-A.

groups are shown in Fig. 11.2 (and section 11-A.4), which is a revision of the pattern shown in Blaha (2021c).

Layers	NORMAL		DARK	
	4	4	4	4
4	SU(2)⊗U(1)⊗SU(3)⊗U(1) 4 Space-time Dimensions	Generation + Layer Groups	SU(2)⊗U(1)⊗SU(3)⊗U(1) 4 Space-time Dimensions	Generation + Layer Groups
4	SU(2)⊗U(1)⊗SU(3)⊗U(1) 4 Space-time Dimensions	Generation + Layer Groups	SU(2)⊗U(1)⊗SU(3)⊗U(1) 4 Space-time Dimensions	Generation + Layer Groups
4	SU(2)⊗U(1)⊗SU(3)⊗U(1) 4 Space-time Dimensions	Generation + Layer Groups	SU(2)⊗U(1)⊗SU(3)⊗U(1) 4 Space-time Dimensions	Generation + Layer Groups
4	SU(2)⊗U(1)⊗SU(3)⊗U(1) 4 Space-time Dimensions	Generation + Layer Groups	SU(2)⊗U(1)⊗SU(3)⊗U(1) 4 Space-time Dimensions	Generation + Layer Groups

Figure 11.2.. Four layers of Internal Symmetry groups in NEWQUeST (omitting Connection Groups). The groups in each layer are independent of those in other layers. The groups in each block of each layer are independent of those in the other blocks. Each block contains 16 dimensions. The block dimensions furnish fundamental representations for the groups listed. The entire set of blocks contains 256 dimensions. Each layer contains 56 internal symmetry dimensions. The first two columns are for the "Normal" sector. The last two columns are for the "Dark" sector (although most of the Normal sector is Dark observationally at

present.) This figure also holds for UST with the addition of U(1) groups. The eight sets of 4 real dimension space-times combine to give a 4 real dimension space-time and seven U(2) Connection groups.

Connection Group Applied to Fermions

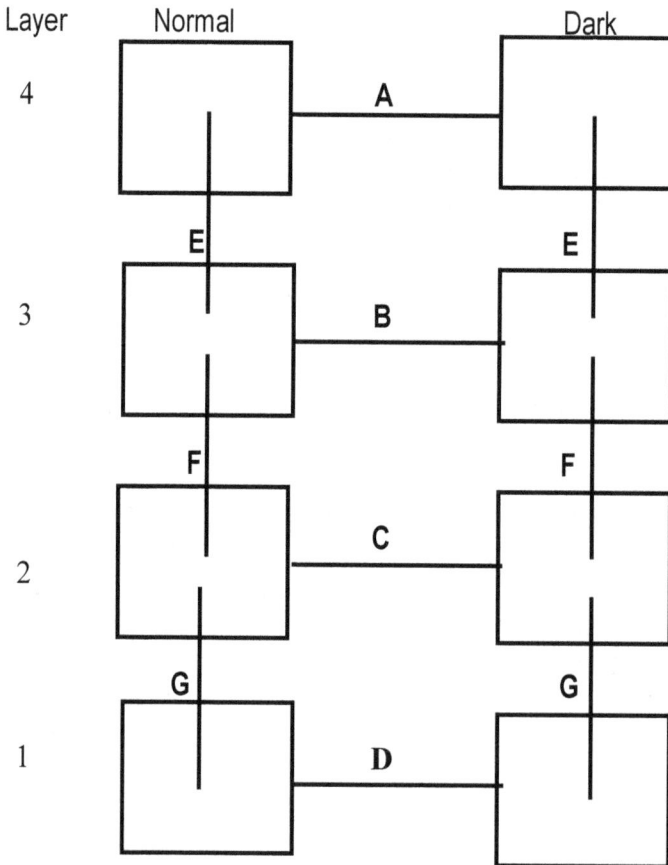

Figure 11.3. The seven U(2) Connection groups (shown as 10 lines)   between   the   eight   NEWQUeST/NEWUST   blocks. Connection groups are obtained by transfering 28 dimensions from QUeST space-time to internal symmetries with the consequent reduction of the space-time from four octonion (complex quaternion) coordinates to four real coordinates. The Connection groups generate rotations and interactions between corresponding fermions and vector bosons of each pair of blocks. See Appendix 11-A. The Normal and Dark sector U(2) vertical connections above (E, F, G) represent the same U(2) groups.

Chapter 15. Possible Relation to SuperString Theories, Composition Algebras

Seven of the ten spaces of Octonion Cosmology appear to correspond to SuperString Theories:

Cayley-Dickson Number n	Space-time Dimension	Spinor Size	Cayley Number = Supercharge	Composition Algebra	SuperString Theory
4	6[13]	8	16	Complex Octonion	Type I
5	8	16	32	Quaternion Octonion	Type 2
6	10	32	64	Octonion Octonion	Heterotic?
7	12	64	128	Complex Octonion Octonion	?
8	14	128	256	Quaternion Octonion Octonion	?
9	16	256	512	Octonion Octonion Octonion	?
10	18	512	1024	Complex Octonion Octonion Octonion	?

The higher Cayley number spaces have yet to be matched to SuperString Theories. ***Their discovery would be of great interest!*** *Then a* possible match between Octonion Cosmology and SuperString theories may emerge.

The ? indicate additional theories that are suggested by the order of type I and type II supercharges, which are Cayley numbers. See Fig. 1.1.

---

[13] Possibly 4 with two compacted dimensions.

Note the type I and type II SuperString Theories are assigned to different Octonion Cosmology spaces.

The comparison of Octonion Cosmology, with its series of steps connecting to NEWQUeST and the Unified SuperStandard Theory (NEWUST), with the goal of SuperString Theories to connect to Elementary Particles suggests direct steps from SuperString Theories to Particle Theory are likely to be difficult. The possibility of SuperString Theories as a basis of Octonion Cosmology is evident.

END OF EXTRACT FROM CHAPTERS 11 & 15 of
Quantum Space Theory

THE TEN OCTONION SPACES SPECTRUM

Octonion Space Number O_s	Cayley-Dickson Number n	Cayley Number d_c	Dimension Array Total d_d	Space-time-Dimension q	Fermion Spinor Array Total d_s
0	10	1024	1024×1024	18	512×512
1	9	512	512×512	16	256×256
2	8	256	256×256	14	128×128
3	7	128	128×128	12	64×64
4	6	64	64×64	10	32×32
5	5	32	32×32	8	16×16
6	**4**	**16**	**16×16**	**6^{14}**	**8×8**
7	3	8	8×8	4	4×4
8	2	4	4×4	2	2×2
9	1	2	2×2	0	1×1

Figure 1.1. The Octonion Cosmology ten space spectrum. The space for our universe, number 6, (Cayley-Dickson number 4) is in bold type.

\

---

[14] The space-time dimensions become 4 as in the Unified SuperStandard Theory (UST) through either transfer of dimensions to internal symmetries or by the compactification of two dimensions. As a result the pattern of fermion-antifermion annihilations producing spaces 7, 8, and 9 is disrupted.

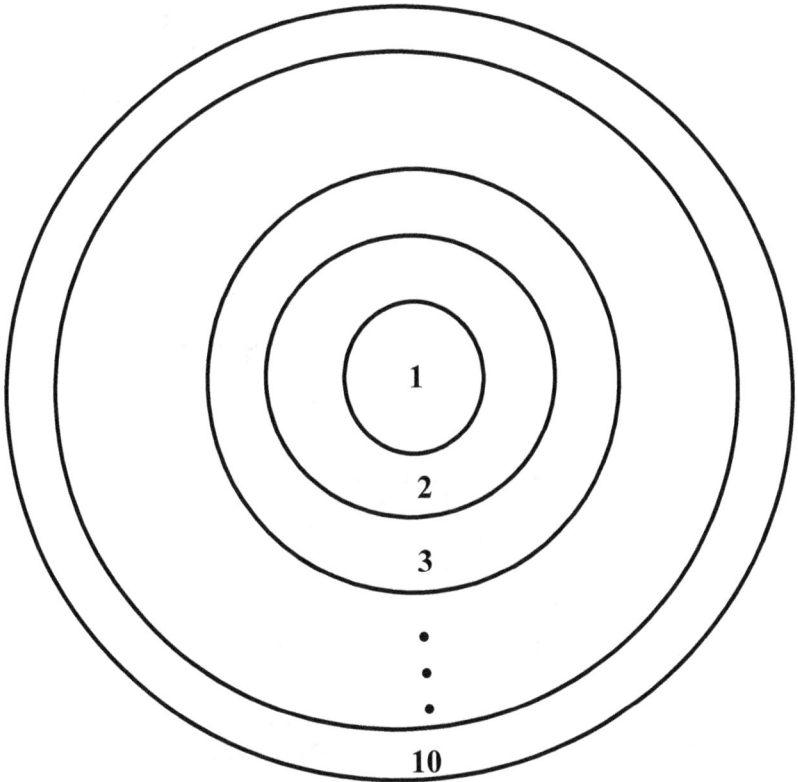

Figure 1.2. Numbered form of the ten space instances of Octonion Cosmology. They are generated by fermion-antifermion annhilations. The outermost space instance (Cayley-Dickson space 10) is generated first from the 20 space-time dimension space. The Space 9 instance is generated within space 10 and so on.

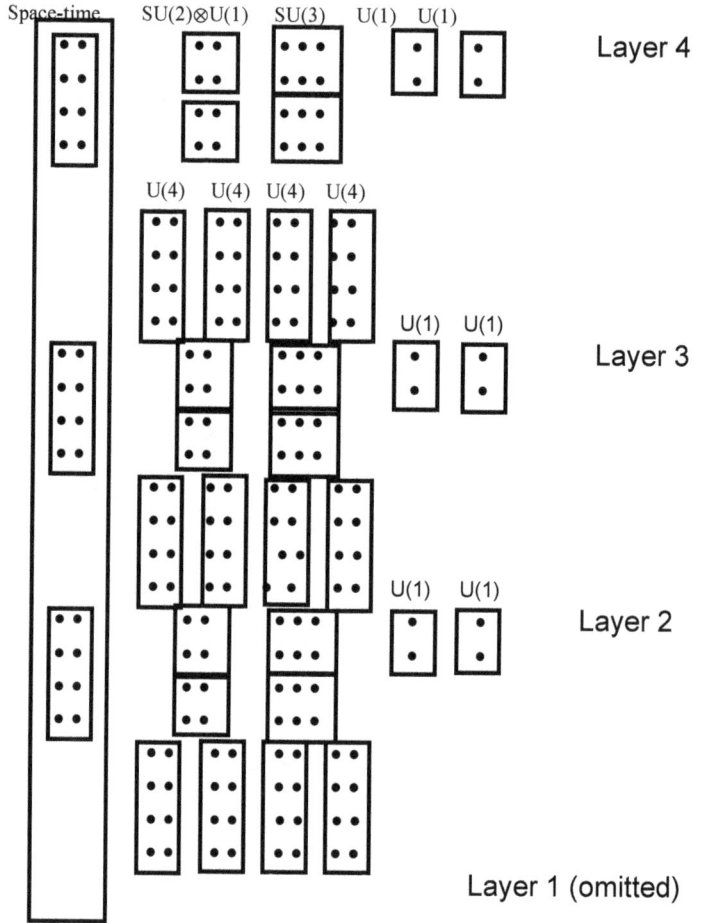

Figure 1.3. Three of the four layers of QUeST internal symmetry groups (and space-time) for Cayley-Dickson space 4. Layer 1 which has an identical form was omitted due to "page space" limitations. Note the left column of blocks combine to specify a 4 dimension octonion space-time. Note each layer has 64 dimensions.

The Fermion Periodic Table

NORMAL FERMIONS | DARK FERMIONS

Layer 4
Generation mixing in the
Four layer Mixing
generations of each species for each

species separately for each layer.
of each species

...

Layer 3

Layer 2

Layer 1 – Our Layer

Figure 1.4. Fermion particle spectrum and partial examples of the pattern of mass mixing of the Generation group and of the Layer group. Unshaded parts are the known fermions with an additional, as yet not found, 4th generation. The lines on the left side (only shown for one layer) display the Generation mixing within each layer. The Generation mixing occurs within each layer using a separate Generation group for each layer. The lines on the right side show Layer group mixing (for Dark matter) with the mixing among all four layers for each of the four generations individually. There are four Layer groups for Normal matter and four Layer groups for Dark matter.. There are 256 fundamental fermions. QUeST and UST have the same fermion spectrum.

A SMALL COSMOS

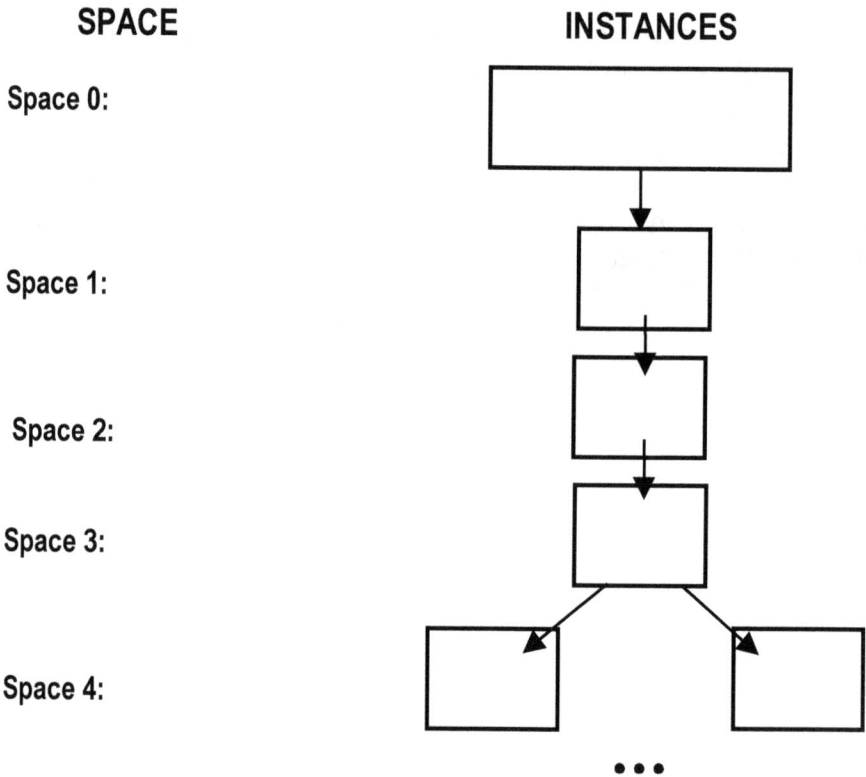

SPACE

INSTANCES

Space 0:

Space 1:

Space 2:

Space 3:

Space 4:

Figure 1.5. A hierarchy of space instances one "sibling" and three "cousin" universes. *The entire hierarchy resides in space 0.*

2. Impact of Octonion Cosmology on Epistomology

The original book *21st Century Natural Philosophy* appears as the third section of this book. It describes the Atomic Period, which was concerned with atoms: "everyday" particles such as leptons, neutrinos and quarks. The new Octonion Cosmological Period began in January, 2020. Its physical theory is outlined in chapter 1. In this chapter we will consider the changes in 21st Century Epistomology wrought by this revolutionary new theory. We will do this by describing the changes from the previous Atomic Period referring the reader to part 3 for additional detail on the corresponding Atomic Period concepts.

2.1 Epistomology

Epistemology is the study of the nature of human knowledge. In the Atomic Period Epistomology was confronted with the problem of understanding the nature of knowledge in the quantum regime.

In the Octonion Cosmological Period one has to consider the epistemology of the Atomic Period since the theories of UST and NEWQUeST describe the quantum theory of elementary particles and atoms. However one also needs to consider the other nine spaces (universes) of Octonion Cosmology. In the case of the other spaces we currently have no concrete knowledge of their nature or their existence.

2.2 Direct Knowledge of the Nine Other Spaces (Universes)

The direct acquisition of knowledge of the nine other spaces beyond the space, within which our universe resides, is not feasible at present or in the immediate future. There is no way to enter the other spaces. There are no known instrumentalities that would allow us to eavesdrop on other spaces. Direct knowledge of the nine spaces may never be possible. Therefore the nine other spaces exist only as theoretical constructs. However there appears to be an indirect method of investigating the other spaces.

2.3 Indirect Knowledge of Other Spaces

Indirect knowledge of the nine other spaces may be possible. As pointed out in section 1.6 previously there is indirect possible evidence of the Megaverse, and possibly higher spaces, from the *form* of the symmetries seen in NEWQUeST and UST:

The form of the square spinor arrays generated by fermion-antifermion annihilation gives 64 dimension blocks and 16 dimension blocks as well as 32 dimension composite blocks that are evidenced in the NEWQUeST (and NEWUTMOST) fermion spectrums and in the internal symmetry group structure.

At present the known symmetries in Nature are SU(2)⊗U(1) – the ElectroWeak interactions and SU(3) – the Strong Interaction. As we saw in chapter 1, UST and NEWQUeST imply the existence of many other symmetries and interactions. The discovery of these, presently unknown, interactions *and of their pattern,* could support UST and NEWQUeST and, more importantly, *imply the existence of the Megaverse (Multiverse) space as well as the spaces above them in the Spectrum of Spaces.*

Thus indirect epistemological knowledge of Octonion Cosmology is possible with answers to traditional questions: Is there any certainty in the knowledge of the spaces (universes)? What is acceptable evidence for their existence? What does knowledge of them imply? What further proof of the existence of these spaces is possible?

These questions may be answerable based on the existence, and organization, of the knowledge in UST and NEWQUeST.

2.4 Octonion Cosmology Uses the Same Words as Standard Model and UST

It is sometimes thought in Epistemology that the nature of knowledge can be reduced to words. Epistemologists define words with words and then contend with conflicting interpretations of the meaning of words.

This viewpoint is particularly salutary for Octonion Cosmology. If one considers the "words" of Nature: matter, energy, interactions, substances, symmetries and so on, then one sees that Octonion Cosmology posits the same reality for each word in our universe with the Standard Model, the Unified SuperStandard Theory, and QUeST; for the words of the Megaverse, and for the words of the nine spaces. Our book *Beyond Octonion Cosmology* shows that the Cosmos is the "onion" that is often talked about. See Fig. 1.2 for the "onion view" of the Cosmos.

Thus the epistemology of the Atomic Period carries over directly to the other nine spaces, which can be called "much more of the same."

The epistemology of the Octonion Cosmologic Period is the same as that of the Atomic Period.

3. Impact of Octonion Cosmology on Metaphysics

As in the Atomic Period case, we shall take Metaphysics to be the study of the contemporary theory of ultimate Reality realizing that our views of Reality have changed drastically in the past hundred odd years. Octonion Cosmology implies another change in Metaphysics.

3.1 Concepts of Metaphysics

Metaphysics is based on our experience of everyday phenomena. It addresses issues relevant to Reality. Questions that it raises include:

1. The nature of matter and its reduction to simpler substances.
2. Modality: The questions of necessity and contingency based on the theoretical concept of a plurality of worlds.
3. Causation – The study of Cause and Effect.
4. The nature of actions and events.
5. The nature of space and time.
6. The nature of Change.

Octonion Cosmology addresses these issues in a manner that often differs from the metaphysics of the Atomic Period and earlier periods.

3.2 The Nature of Matter in Octonion Cosmology

The nature of matter is extended in Octonion Cosmology from that of our universe (the Standard Model, Unified SuperStandard Theory

(UST), and QUeST) to that of the other nine spaces. Many more types of matter (with differing internal quantum numbers) appear in higher spaces. However all types of matter can still be viewed as manifestations of the same substance. The substance can best be described as "nothingness structured by symmetries and conservation laws." Matter can be described by quantum wave functions in all cases.

Octonion Cosmology introduces a new type of matter in its Quantum Space Theory formulation. It defines quantum space fields, which are particle-like entities with mass and momentum. A space field also contains an internal space with an array of dimensions. This set of dimensions has some dimensions that support a space-time. Its other dimensions define subspaces (fundamental representations) of internal symmetries such as ElectroWeak symmetries. There is energy and particles within the internal space.

Octonion Cosmology views our universe as such an internal space residing within a space particle in the Megaverse (Multiverse).

Space particles are a new type of matter much beyond the particles of the Standard Model of Elementary Particles. They have mass and momentum, and move within the Megaverse. Space particles can interact, can collide, and behave like ordinary particles in most ways. See Fig. A in chapter 1.

3.2.1 Space Particles and Substances

Metaphysics concerns itself with the nature of matter. Matter is known to consist of atoms, which are, in turn, composed of elementary particles. As pointed out above, ordinary matter consists of a single substance, which is a structured nothingness. Energy is similar since all types of energy are composed of particles: photons and so on.

Space particles are also structured nothingness. Admittedly they have additional structuring. But still they remain structured nothingness.

Thus the Metaphysics of matter in all its forms leads to the same substance. One might add the vacuum to the structured nothingness form of substance since we can view its particles in the Fermi Sea, and its energy, as structured nothingness.

3.2.2 The Dimensions of Octonion Cosmology

In the Atomic Period dimensions were mostly thought to be of significance for space and time although the dimensions of irreducible representations of internal symmetry groups were known and used.

In Octonion Cosmology dimensions are of greater significance. First they are specified by the spinor arrays of colliding fermion-antifermion pairs, which transform to the dimensions of internal spaces of generated space particles. The dimensions of an internal space play two roles. One set of dimensions are space-time dimensions such as the four dimensions of our universe. The remainder of the set of dimensions becomes the dimensions of irreducible representations of internal symmetry groups. The elementary particles within the internal space are in representations of these internal symmetry groups.

Thus dimensions play a major role in Octonion Cosmology. *Octonion Cosmology unites space-time dimensions with internal symmetry dimensions.*

Since spaces are generated by fermion-antifermion annihilation one might say that matter generates dimensions and thus spaces. The generated dimensions characterize particles. Particles only exist by virtue of their internal symmetries and the conservation laws that they embody.

Thus one might also say matter is "generated" by the properties of dimensions and their associated groups.

As a result the well-known mass – energy relation of Einstein can be extended to be a mass – energy – space dimensions relation as pictured in section 1.6 with the figure:

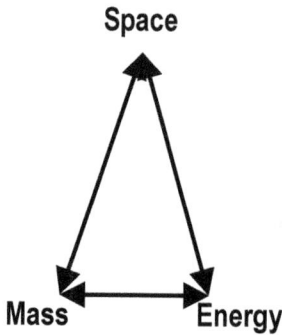

Space

Mass   **Energy**

Figure B. The golden triangle of mass, energy, and space.

Mass can generate space dimensions. Space dimensions support massive particles and energies.

3.3 Causation – The study of Cause and Effect

The study of Causation in the Atomic Period and the Octonion Cosmologic Period devolves to the quantum level. Causation is reduced to the interactions of particles and energy, which are described by Lagrangian models and theories. Thus it is no longer a field of inquiry except for one concern: What is the Prime Mover, if there is one, that started the Cosmos? In Octonion Cosmology the question is "Who/What instantiated the twenty space-time dimension space described in chapter 1." This space is the origin of the Cosmos.

3.4 The Nature of Actions and Events

Actions and events are generated by dynamics specified by Lagrangians in the Atomic Period and the Octonion Cosmologic Period. They only differ in the vastly larger number of interactions in Octonion Cosmology.

3.5 The Nature of Space and Time

Space and time are viewed differently in Octonion Cosmology. In the Atomic Period they were conceptually separate but interrelated by the Theory of Relativity. They are completely independent of Internal Symmetries such as SU(3). In Octonion Cosmology they retain this relationship due to the Theory of Relativity.

However, in Octonion Cosmology space-time is "united" with internal symmetries. For example, four dimensions are allocated to space-time coordinates in our universe.

The other dimensions of a space are allocated as coordinates for internal symmetry subspaces (fundamental group representations.) For QUeST (our universe) 252 dimensions are allocated for internal symmetry subspaces.

Conceptually space-time is on the same footing as internal symmetry dimensions. Their "Reality" is the same. Their "substance" is the same. In a sense, our human view of Reality requires expansion.

3.6 The Nature of Change

Change is always with us. In the Atomic Period and the Octonion Cosmologic Period Change is the same—ultimately the consequence of Quantum Theory, which is similar in both Periods.

In Octonion Cosmology we are required to enhance Quantum Theory at the level of very high energies (short distances) because of the higher dimension space-times of the higher spaces (universes). Space 10 for example has 18 space-time dimensions.

In higher dimension space-times we must introduce a new short distance quantum effect that enables quantum field theory to avoid a plethora of divergences (blow-ups). This new quantum feature is called Two-Tier Quantum Theory and is described in the author's 2005 book *Quantum Theory of the Third Kind*. Thus Octonion Cosmology requires a changed Quantum Theory.

Change has other unforeseen aspects due to Quantum uncertainties and probabilities. We discuss these aspects later in the part on Predestination and Free Will.

3.7 Coordinate Structure of Quantum Theory

In the Atomic Period we found that position and momentum satisfied an Uncertainty Principle: the more accurate the measurement of position, the less accurate the measurement of momentum and vice versa. The Uncertainty Principle led to Quantum Mechanics and thence to Quantum Field Theory. It was noted that infinities would have emerged in dynamical calculations (such as the calculation of the energy levels of the hydrogen atom) if the Uncertainty Principle did not hold.

In Quantum Field Theory early calculations led to infinities also. These infinities were handled through mathematical techniques called

Renormalization Theory that buried infinities in various (infinite) constants. Thus "simple" theories such as Quantum Electrodynamics and ElectroWeak Theory were able to obtain results without manifest infinities.

This author suggested that an addition to the Uncertainty Principle should be made in his 2005 book[15] *Quantum Theory of the Third Kind*. He suggested that the Uncertainty Principle should include uncertainty in space and time coordinates that only appeared at ultra-high energies. This extension of the nature of coordinates was called Two-Tier Coordinates since it introduced a second level of uncertainty.

Two-Tier Coordinates, when used in Quantum Field Theory calculations in any model such as ElectroWeak Theory and so on, eliminated *all* infinities including those of fermion triangle diagrams.[16] No need for renormalization group techniques to eliminate infinities.

Two-Tier Coordinates is very necessary in Octonion Cosmology because it has spaces with space-times with more than four dimensions. Without it, calculations could not be made in these larger space-times without uncorrectable infinities.

Thus Octonion Cosmology forces us to use an extension of the Uncertainty Principle to include Two-Tier Coordinates to avoid infinities in space-times with greater than four dimensions. There are six spaces of this sort. See Fig. 1.1.

---

[15] And actually earlier in the 2000's.
[16] It had the advantage that "low" energy calculation results in Quantum Field Theory with Two-Tier Coordinates were unchanged.

The Octonion Cosmologic Period requires a major extension of the Uncertainty Principle whose effects can only be seen in experiments at ultra-high energies beyond those currently accessible.

3.8 Spinor Spaces

An examination of the ten octonion spaces (universes) and the 20 space-time dimension space reveals that they all originate as spinor spaces – spaces generated by the spin of space fermions and bosons. The spinor spaces are recognized to be "true" spaces (universes) when one examines the scalar space boson wave function. Then one sees that the creation and annihilation operators within the wave function have spinor indices that "make" them resemble operators in a space. This resemblance is enhanced by the Dirac-Good symmetry group that supports transformations among the operators just as elementary particles undergo SU(3) and SU(2)⊗U(1) transformations.

As a result spinor spaces[17] assume their true form as spaces (universes) capable of supporting particles, energy, space-time coordinates, and internal symmetries. Physically true spaces!

Spinor spaces were not evident in the Atomic Period. They blossom in the Octonion Cosmologic Period.

3.9 Possible SuperString Relation

SuperString theories may be related to the Octonion spectrum of spaces. The table in the chapter 15 Extract in chapter 1 shows that known SuperString theories have supercharges that mirror the Cayley numbers of the Octonion Spectrum. It is possible that the octonion space spectrum

---

[17] Spinor spaces have been considered before for Twistor Theories by Penrose and associates. They are conceptually different from spinor spaces discussed here.

emerges from a deeper SuperString formulation. The proof of this connection would require the discovery of four additional SuperString theories.

3.10 The Trend of theories

The author's theories leading to Octonion Cosmology show an intriguing trend towards deeper, more abstract foundations. They may be diagrammed as

NEWUST $\supset$ NEWQUeST $\supset$ Octonion Space 4 $\supset$? SuperString Theory A

where the symbol $\supset$ can be read as "is contained in". The Megaverse (Multiverse) also has a progression of theories:

NEWUTMOST $\supset$ Octonion Space 5 $\supset$? SuperString Theory B

where A and B designate SuperString theories.

A question mark is placed after $\supset$ above since the connection to SuperString theories remains to be demonstrated.

These chains of theories are a new feature that is present only in the Octonion Cosmologic Period. They reflect an interesting point of advancement in fundamental Physics theories.

3.11 Reality vs. Theory

The concepts of Reality in the Octonion Cosmologic Period do not change from those of the Atomic Period. However they are deepened in the sense that the Reality of the NEWUST theory, which is that of the Atomic Period, now follows from the Realities of the spaces above space 4 of the Octonion spectrum. Most of the realities of the higher spaces (universes) are related by the fermion-antifermion annihilations that lead to space 4. The form of their Internal Symmetries carries over to our Reality. We see that in the discussions of section 2.3 and the Extract in chapter 1.

Reality thus becomes a many-faceted construction that follows from the overall structure of Octonion Cosmology.

PART II: A Quantum Basis for Predestination, Fate, and Free Will

4. A Quantum Basis for Predestination, Fate, Free Will

Fixed Fate, free will, foreknowledge absolute.

J. Milton, *Paradise Lost*, Book ii, line 560

Predestination, Fate and Free Will are traditionally viewed as subjects for Moral Philosophy and Theology. However, the advent of Modern Physics, particularly Quantum Theory, has made these topics susceptible to discussion as part of Natural Philosophy. In this chapter we formulate a Quantum Natural Philosophy view, which leads to a resolution of their often supposed conflicts.

4.1 Source of the Predestination/Fate/Free Will Controversy

Predestination and Fate are troubling concepts that are accepted in some religions and rejected by others. They state that all events happen through the design of God or Fate, and that free will does not truly exist—free will is viewed as an illusion. Believers in Free Will usually discount Predestination and Fate as, at best, illusory and often take an agnostic or atheistic position. They tend to believe that if God did not exist, then Predestination would not exist. Fate would also be a matter of accidentals. *The existence and features of God are the major source of the controversies over Predestination, Fate and Free Will.*

In this chapter[18] we suggest that Quantum Theory can provide an understanding of Predestination and Fate that enables Free Will to also exist. God will be shown to be able to intervene in the Cosmos without violating Physical Laws in a manner based on Quantum Theory.

4.2 God's Tape of Fate of the Cosmos and Predestination

Supposing God exists, we have suggested[19] the possible Predestination view that God, in effect, has an infinite tape, the "Tape of Inexorable Fate," containing all aspects of the Cosmos from the remote past to the remote future. His foreknowledge indicates that He is fully aware of the entire contents of the tape everlastingly in the past and in the future.[20,21]

God's "foreknowledge absolute" is the predestination to which many religions refer. It appears to eliminate the possibility of arbitrary intervention by God into Cosmic events at selected points in a manner not present on the tape. Yet God's intervention does occur according to their religious thought, and any intervention by God must appear in the Tape of Fate, since it is a faithful record of eternity.

---

[18] This chapter is adapted from Blaha (2018e). God will be assumed to exist. God will be assumed to act consistently with the Physical Laws that He created. God will be assumed to not violate Physical Laws. Consequently the actions of the Godhead are susceptible to *Physical* analysis.

[19] Blaha (2018e).

[20] Time travel, which appears to be possible in principle, can be accommodated within the Tape of Fate as simply a flattening of the tape segments for a time travel trip to one composite tape segment—a multiplexed tape segment.

[21] Western religions posit God is totally cognoscenti of the entire life of the Cosmos.

God cannot arbitrarily and randomly intervene in Cosmic events because God's Physical Laws[22] cannot be violated—*superficially* contradicting God's assumed limitless power. A moment's thought shows that all God's interventions at various times in the "infinite life" of the Cosmos must have existed as a reality forever—from the infinite past to the infinite future of the Cosmos. Thus we have the result: intervention without violation of Physical Laws.

4.3 The Need for a Quantum Cosmos

In Blaha (2018a), and earlier, we suggested that Quantum Theory was the simplest way to ensure that matter and energy were both made of discrete particles.[23]

We now consider another, perhaps deeper, justification for Quantum Theory. While Quantum Theory has many facets, one facet of great importance in our view, is that it introduces chance into physical processes. When an event takes place, such as the collision of two particles, there will be varying results of the collision. Each possible result has a certain probability of happening—just as flipping a coin has a probability of "coming up" heads and a probability of "coming up" tails.

4.3.1 God's Intervention in the Cosmos

The introduction of chance in physical events opens a "new" possibility for God's intervention in Cosmic events. Without chance God's intervention would be strictly limited by Physical Law (which we

---

[22] God's laws are created and required as the "Unmoved Mover." They cannot be violated. (All Physical experience confirms this point.) They were created by God!

[23] Quantum effects, because they are probabilistic in nature, can be selectively used by God to achieve God's "goals."

assume God established absolutely) making "miracles" very limited, and in most cases impossible.

With chance in physical events, it becomes possible for God to "tweak" an event quantum mechanically[24] to make a certain result happen, without violating Physical Law, and consistent with the quantum mechanical probabilities. The tweak can then cause, or be accompanied by, a series of further God-manipulated quantum events.

The result is a cascading set of events caused by Quantum tweaks consistent with Quantum probability, and Physical Law—a chain reaction of events that can generate a miracle or even change history. One remembers here the lament of Richard III of England, "My kingdom for a horse." (We will not comment on the chain of events that would have put a horse within Richard's grasp.)

Thus God can intervene in events without violating Physical Law through quantum effects.

4.3.2 Quantum Source of Free Will

Free Will is usually viewed from the aspect of a person's mental state. However, free will has several forms. Some forms relate to the person's perceived situation with free will exercised to maximize the person's benefit and/or views. This form involves a significant part of the person's consciousness.

Another form of Free Will is chance decisions. A good example of this form is spot decisions as to where to place chips on a roulette table. This type of decision can be based on guessing and may reflect a chance event in a person's mind. *Free will based on chance, miniscule*

---

[24] God selects one of the possible probabilistic outcomes to suit God's purpose. This selection process was known to God in the Beginning.

events in a person's mind can be the result of cascading Quantum tweaks.

Thus a person can have free will without God's intervention. But God can also intervene by making Quantum choices of outcomes in a person's neurons.

God can thus intervene in the Cosmos to implement change, guide free will, and execute miracles *through Quantum effects*. On occasion God may load the dice. But as Einstein suggested, "Not maliciously."

4.4 Quantum Observer Problem

Having brought in Quantum Theory to support intervention by God we must now consider an important Quantum Theory problem that is resolved by the existence of God.

The problem is the need for a Quantum Observer to view an event and to observe which of the possible results actually occurs in reality. A famous example is Schrödinger's Cat. A cat is placed in a box with a radioactive element that decays after a "random" time. The setup in the box causes the cat to die when the radioactive element decays. The problem is: After x minutes is the cat alive or dead? The usual answer is it can only be found if the box is opened by an observer. Before that point in time the cat is presumably in an unknown state with the reality of its life in doubt.

Thus Reality becomes an issue of observation. Some have suggested that complete knowledge of the entire element-cat-box apparatus as a quantum state would resolve the issue. But that is unlikely because the composite system is quantum and subject to chance as well.

One could extend the apparatus state to the entire universe. But that would still be quantum issue.

So the tenure of the cat's lifetime remains uncertain forever. A sad situation it appears.

The situation of Schrödinger's Cat applies in many processes throughout the Cosmos. As a result it appears that Reality is an ever present issue for the Cosmos.

4.5 God as the Ultimate Quantum Observer

The only clear-cut solution to the Reality problem is God—the Ultimate Quantum Observer.[25] God need only look at "the Tape of Fate" to see the results of all Quantum events. Thus God gives Reality to the Cosmos. An infinite progression of observers is thus avoided. There is always an implicit observer.

4.6 Events Without Observers

A question raised in some religions: does an acorn make a sound when it falls to the ground if no one is present to hear the sound, is also resolved. God, the Unmoved Mover, is the omniscient Observer in all quantum and non-quantum events. It makes Reality real whether or not a human observer is present.

---

[25] Recently generalizations of the Schrödinger's Cat "experiment" have been considered in which an observer is placed in the box. Then it appears that if two observers, outside the box, open the box they will find different observations making a quantum paradox. This controversial result has been the source of much discussion..Again, our choice of God as the Ultimate Quantum Observer eliminates the paradox. See D. Frauchiger and R. Renner, arXiv:1710.05033v1 (2017) and arXiv:1604.07422v1 (2016) for detailed discussions of the paradox.

4.7 God Knows All at All Times

All eternity is predestined since all events are known to God simultaneously at all times. Yet Quantum tweaks (prespecified in the Beginning) make it possible for God to make events happen without violating the Physical Laws that God established. *All* events, including tweaks, are recorded on the Tape of Fate *forever* from the "Beginning" (which is not truly a beginning since there is no beginning) to the end of eternity (which is not truly an end since there is no end).

Predestination, Fate and Free Will are reconciled through Quantum Theory. The Observer issue of Reality is resolved by the existence of God. *Some will consider the need for a universal Observer as the ultimate Physical proof of the existence of God.*

PART III: 21st CENTURY NATURAL PHILOSOPHY

21ˢᵗ CENTURY NATURAL PHILOSOPHY OF ULTIMATE PHYSICAL REALITY

STEPHEN BLAHA, PH.D.

MCMANN-FISHER PUBLISHING
MMVIII

ISBN: 978-0-9819049-9-3

Cover Credits
Cover by Stephen Blaha © 2010. All rights reserved.

rev. 00/00/01   January 5, 2011

PREFACE

Western Philosophy seems to have originated in the 6th century BC. Thales of Miletus is considered by many to be the first Greek philosopher. The core of his philosophy was a strictly natural description of the world (universe) based on the assumption that water was the primary, and only, fundamental substance composing the universe although it manifested itself in numerous forms that appeared to be different substances. The basis of his philosophy was deductions based on the close observation of nature.

In this work we will return to the monistic ideas of Thales (and his successors such as Anaximander and Anaximenes of Miletus) and show that a much closer observation of nature at the sub-quantum level leads directly to the view that the universe is composed of one substance that can assume a variety of forms that we call elementary particles and can also have a "lack" of form (by being an infinite superposition of all forms) that we call the vacuum. The substance is not water or air as the 6th century philosophers proposed, but does bear comparison with the *apeiron* of Anaximander in that it is best described as "infinite" and without "distinctive" features where "infinite" means creatable without limit and "distinctive" means not describable in terms of everyday experience.

The natural philosophy presented in this volume is based on the general nature of quantum field theory, and particularly on a detailed mathematical theory of Logic and Elementary Particle theory (The Standard Model) presented in Blaha (2010c). We take the position stated in Blaha (2010c):

> As science evolves it is becoming increasingly clear that all events and phenomena in our universe ultimately follow from fundamental physical Reality. Some phenomena, such as the processing of abstract concepts in the human mind, are very distantly related to fundamental physical Reality, and it may be difficult (and perhaps presently impossible) to derive a complex phenomenon from fundamental physical Reality. Nevertheless the connection exists despite our inability to derive it. So we regard all things (except the spiritual) as emanating from physical Reality. Our present inability to derive connections is a matter of happenstance; but the principle of a unitary Reality based on fundamental physical principles has the force of historical trends behind it as the developments in the twentieth century clearly show.

Thus all of physical Reality is based on the fundamental substance, on the forms it can assume, and on aggregates of these forms which we call matter and energy. In expressing this view we exclude theological and spiritual matters from consideration. This work is strictly about the material world.

In the interests of reaching a wide audience, there is no mathematics in this work. Those wishing to see mathematical detail are directed to Blaha (2010c), and earlier work.

1. What is 21ˢᵗ Century Natural Philosophy?

Philosophy is a wide area of inquiry with many subdivisions. Broadly speaking, it is the careful study of the fundamental basis of our beliefs. In earlier times Philosophy[26] and Physics were closely intertwined and this joint study of the fundamental nature of the physical world was called Natural Philosophy or Metaphysics.

Until recent centuries, observations and theories of Physics were a driving force for developments in Natural Philosophy (Metaphysics) and, indeed, in the Arts. Alexander Pope's great lines,

Nature and Nature's laws lay hid in night;
God said, Let Newton be! and all was Light

illustrates the profound effect of Physics on the Arts as well as Philosophy.

With the enormous growth in our knowledge of the physical world in the 19ᵗʰ and 20ᵗʰ centuries Physics became separated from Metaphysics and Philosophy. An important cause of this divergence was the emergence of Mathematics as the language of Physics while Philosophy, and Metaphysics in particular, continued to be largely based on human languages such as English, German and French. In addition, philosophers and metaphysicians tended to seek a generality that led to conceptual developments that could scarcely be considered relevant to modern Physics.

Thus the divergence beginning in the early 19ᵗʰ century led effectively to a divorce between Philosophy and Physics. Lip service was paid by both "parties" to the value of the other party's investigations. But a creative relationship that promoted advances in both areas of study was lacking.

We propose a new Natural Philosophy based on the revolutionary new Physics that has emerged in the past two centuries. More precisely, we propose to take the qualitative aspects of the most fundamental observations of Physics in the 20ᵗʰ century and thereupon erect the edifice of a 21ˢᵗ century Natural Philosophy. As we mentioned in the Preface it has become clear that all non-spiritual phenomena are

---

[26] Again we note that we will exclude Theological Philosophy from our considerations.

ultimately physical in nature including phenomena of the mind and consciousness.[27] Thus the key to all natural phenomena lies ultimately in the fundamental physical theory of Nature, often called "The Theory of Everything" or the fundamental theory of Physics.

    Currently The Standard Model of Elementary Particles and the Theory of General Relativity are the only known relatively complete and experimentally confirmed fundamental theories of Nature. There are details that remain to be understood in both, and major questions to be answered, originating in cosmology, such as the nature of Dark Energy and of Dark Matter. However all known phenomena seems to ultimately follow from The Standard Model and General Relativity. Thus we will base our development of 21ˢᵗ century Natural Philosophy on what we have learned from the experimental successes of The Standard Model and General Relativity and leave open questions to future inquiry.

    Science is generally concerned with the description of natural phenomena and its success is based on its ability to predict the outcome of experiments and natural processes. Thus it answers descriptive "How" questions and does not answer explanatory "Why" questions. When it appears to answer "Why" questions it really is relating some phenomena to a more fundamental level of description, and thus is only superficially a "Why" answer. The basic nature of a "Why" answer requires the scientist to exclude all alternate possible answers in all possible universes. The current state of our knowledge of physical Reality, and our understanding of possible alternate realities, does not enable us to achieve such a result. Nevertheless it appears possible to develop a provisional set of principles that can enable a theoretical investigation of the basis (the "Why") of our Reality, and of alternate realities (universes), to begin.

---

[27] But not necessarily the thoughts of the mind.

2. Epistemology

This chapter begins a study that leads to a qualitative, contemporary understanding of physical Reality at its most fundamental level – The Standard Model of Elementary Particles combined with the General Theory of Relativity.[28] Some primary issues that we will examine in the next few chapters with a view towards our goals are Epistemology – the nature, scope and limitations of human knowledge; Metaphysics – the nature of ultimate Reality; and Logic – the true relationships between the parts of ultimate Reality. The author believes that the impact of Quantum Theory and the Special Theory of Relativity on these areas of study has yet to be fully appreciated. One of our goals is to provoke a thorough reanalysis of these subjects.

2.1 Epistemology

Epistemology is the study of the nature of human knowledge. Since our goal is to understand one specific area of knowledge – physical Reality – our approach to epistemology will be limited to the nature, scope, and limitations of human knowledge of physical Reality. Physical Reality can be viewed as consisting of three parts: cosmological physical Reality (the universe in the large), everyday earthly (planetary) physical Reality, and quantum physical Reality (Reality at the quantum level on the very smallest distance scales). These parts of Reality conveniently break up into distance scales: very large distances – cosmological distances; the distances and sizes of everyday experience – ordinary distances; and very small distances – quantum scale distances. Each of these areas of human knowledge has a different epistemology. Almost all work in the epistemology of knowledge of physical Reality has been at the level of everyday earthly Reality.

Before turning to a study of the epistemology of very large-scale and very small-scale knowledge we should understand the makeup of epistemological theory. One way to begin is to consider an isolated human mind with absolutely no sensory inputs, and then to consider

---

[28] It is clear that the current "fundamental" level is either an interim level preliminary to a deeper level or is only part of a broader fundamental level. The major evidence for this view is Dark Matter and Dark Energy – phenomena that have appeared in Cosmological attempts to account for observed astronomical data. Nevertheless the Standard Model of Elementary Particles has repeatedly been proved to be an accurate description of known earthly experimental data and most cosmological data.

what physical senses are available and how their capabilities and limitations affect the ability of the human mind to acquire knowledge and process it.

So let us first consider a human mind (or any intelligent mind) in *total* isolation with no prior, or current, sensory input and ask what sort of knowledge it can obtain. First it is clear that any knowledge that it might have would be the result of imagination.[29] Conceivably the mind could form some concepts. Perhaps the simplest concept would be number. But not having anything to count, why should it conceive of number.[30] Similarly, all other concepts derive from sensory perceptions and so it is difficult to believe that any concepts would occur to the isolated mind except possibly concepts/ideas of a genetic origin – if such exist.

But there is a difficulty in the mind itself. Modern research has established that sensory inputs affect the wiring of the human mind – as the mind processes the inputs. So a totally isolated mind could be viewed as an "unwired" mind. The thought processes that would take place in such a mind would appear to be haphazard. Thus we are led to conclude that an isolated mind without sensory input and without inherent genetic information would in general have no knowledge and be incapable of thought. The young Helen Keller, who lacked hearing and sight but had a sense of touch, is an (imperfect) example of an isolated mind. Ms. Keller was not rational until after much schooling.

If we now consider the human mind with its usual physical senses: sight, hearing, smell and touch, then a totally different view emerges. For the senses, by providing input to the mind, gives it items to consider, behavior to analyze, and motives (self-preservation, and the acquisition of wealth and power) for survival.

2.2 Origin of Human Knowledge of Physical Reality

The nature of human knowledge of physical Reality at the everyday level of experience has been a subject of discussion for thousands of years. We will not enter to that still ongoing discussion other than to say that everyday knowledge is largely based on direct observation using the human senses of physical Reality.

Rather we will examine the nature of human knowledge at the quantum and cosmological levels. In both cases the nature of human knowledge is observation of physical Reality through instrumentation of

---

[29] The possibility of inherent genetic "knowledge" present in the mind (such as Jungian archetypes) is not excluded.

[30] Even for people who have sensory input the concept of number is not easy. For example members of some tribes cannot count beyond 1 or 2.

varying degrees of complexity, which is interpreted using a theoretical analysis that can be quite complex and often uses probabilistic interpretations. As a result the nature of human knowledge in these areas of Reality is indirect. Theoretical analyses stand between our knowledge of Reality and experimental results. A change in the analysis of an experiment usually leads to a different "knowledge" of the corresponding part of Reality.

2.3 Scope of Human Knowledge of Physical Reality

Thus the scope of knowledge of Reality is in a sense limited by our ability to analyze what we are observing.[31] And so quantum and cosmological knowledge of Reality has a layer of theory between knowledge and observational data. In some areas of science such as elementary particle physics the theoretical analysis of the data is extremely complex and problematic. As our ability to do experiments to determine features of physical Reality becomes greater the theoretical analysis will correspondingly become more critical.

2.4 Nature of Human Knowledge of Physical Reality

In the three areas of knowledge of physical Reality we see patterns or structures of knowledge. First one has a set of facts relating to some aspects of physical Reality. Then these facts are usually organized into a theory based on principles or axioms that presumably are "deeper" then the amalgamation of facts – from which the facts may be derived. Lastly, in an increasingly larger number of theories of various aspects of physical Reality, the theory is put into a mathematical form. Euclidean geometry is considered the example par excellence of a mathematized theory deduced from Reality. Many theoretical areas of physical Reality, emulating Euclid, espouse the goal of becoming a mathematical-deductive theory – a theory with a set of fundamental axioms from which theorems can be derived that describe all phenomena in the area.

Presumably all aspects of physical Reality follow from a fundamental Theory of Everything.[32] But the connections between the

---

[31] Seeing is based on prior knowledge, and fitting observation within the framework of prior knowledge.

[32] Some thinkers have recently asserted that a theory of everything is not possible. This assertion is not true if one believes that ultimately the understanding of each area of physical Reality can be reduced to a physical theory (a set of mathematical or logical rules) that describe phenomena in that area correctly. If one does not accept that assertion then one denies the basis of scientific thought for the past 400 years. Accepting a correct theory of each area of physical Reality exists, one can construct a theory of everything by joining all these experimentally verified areas in one composite theory. The composite theory may be unesthetic and cumbersome but it would be a theory of everything. If the objection is raised that parts of the composite theory might conflict, then the objection is refuted by noting that any parts of the composite theory that overlap must be

various specialized theories is often hard to derive. The Theory of Everything is expected by many scientists to follow from a unified theory of fundamental particles and gravitation. But the derivation of psychological phenomena, for example, from the Theory of Everything is a long and tortuous process that has many gaps currently – primarily at the level of the understanding of the chemistry of the human brain and its relation to psychological states.

The concept of a Theory of Everything makes many of the questions addressed by traditional epistemology pointless. Some of the traditional questions are: Is there any certainty in the knowledge of the universe? What is acceptable evidence for a theory? What does knowing something mean? What is proof? These questions are answered by the existence, and organization, of the knowledge in an experimentally verified Theory of Everything. Beyond that we are reduced to quibbles over points that have no practical significance with respect to understanding Reality.

In effect the reduction of our experimentally obtained knowledge of Reality to a mathematical-deductive system defined with primitive terms is the only reasonable description of Reality. Section 2.7 discusses the essential nature of primitive terms – terms that are not unambiguous but can be made unambiguous by identification with specific entities of physical Reality.

2.5 Limitations on Human Knowledge of Physical Reality

The sources of human knowledge are observation, direct or indirect, and rational analysis of the facts acquired through observation. To obtain factual knowledge about Reality we must be able to obtain information by direct observation using our senses or through instruments that play the role of extending the senses. For example, we cannot see ultraviolet light with our eyes but we can devise an instrument that can "see" ultraviolet light.

When everyday phenomena were the stuff of knowledge, it appeared that we could have a total knowledge of physical Reality. However the situation changed dramatically in the twentieth century. First quantum theory showed us that certain combinations of facts could not simultaneously be ascertained at the quantum level (small distances). For example the position and momentum of an electron cannot be exactly determined simultaneously due to the Uncertainty Principle.

---

in agreement since an experiment in the overlap area can give only one result showing one part to be true and the other part false – contrary to the assumption that all parts are correct. Thus a theory of everything must exist.

In addition, more recently, it has become clear that the amount of matter and the amount of energy in the cosmos are different from that determined by astronomical observations. There is an "unseen" part of the universe. All of our senses and instruments for observation are based ultimately on the four known forces of nature: gravity, electromagnetism, the weak interactions and the strong interaction. The latter two forces are only detectable at high energies (short "quantum-scale" distances). The unseen matter in the universe, Dark Matter (about eighty-five per cent the total matter in the universe) is only known from its gravitational effects at the present time. It seems to be undetectable through the stronger forces of nature although it may be detectable through the Weak interactions. That being the case, Dark Matter can "interpenetrate" (occupy the same space) as the ordinary matter that we see, and form its own unseen part of the universe.

Similarly we see the effects of Dark Energy[33] in the cosmos – particularly in the rate of expansion of the universe. Here again we see the effects but do not see the entity. Our senses, even our senses extended through instrumentation cannot currently acquire detailed knowledge of most of the universe.

Everyday epistemology and metaphysics did not anticipate, and has not explained, the recently found, unseen, and currently unseeable aspects of the universe. Therefore any claims that contemporary epistemology and metaphysics explicate human knowledge or ultimate Reality are incorrect.

2.6 Mind-Body "Problem", and Reality

A topic of interest for many years has been the "Mind-Body" problem. Many people see a discrepancy between the intellectual processes taking place in the mind and the qualities of the human body—particularly the human brain. However, recent research on the human brain using new techniques such as MRI have revealed that specific parts of the brain handle input from the senses, emotions, and even certain thought processes such as number processing (mathematics). We are therefore developing an understanding of the relation between the mind and consciousness, and the physical-chemical structure of the human brain.[34] In a very real sense the human mind is now being seen as a very

---

[33] Energy of unknown origin constituting about seventy-five per cent of the total energy in the universe.

[34] Scientists using fMRI (functional Magnetic Resonance Imaging) have recently found signatures of consciousness. They have detected coordinated activity across the entire brain and conscious processing of images in the mind. See A. Schurger et al, Science, DOI: 10.1126/Science .1180029 (November, 2009).

complex computer, much beyond our biggest and fastest electronic computers, yet still a computing apparatus. Mathematical networks have been developed that simulate aspects of learning, thinking and memory.[35]

If all the aspects of the mind can be reduced to mechanisms of the human brain then the major remaining issue is the nature of consciousness. Consciousness is a conglomeration of the features of the mind that give us a coherent "picture" with which we can observe, process, and respond to external and internal events. Consciousness also generally has a continuity of transitions from one state to another. Perhaps the best simple description of consciousness is a self-controlled computer or television set that undergoes transitions based on internal processing of internal states and external sensory input.

Reality – Senses – Mind – Theory

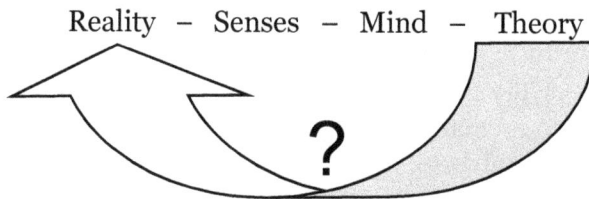

?

Figure 2.1. Why do theories created by the human mind so closely mirror Reality? The author believes it is due to Reality being "derived" from Logic, the mechanism of thought. This topic will be discussed in detail later.

We now have a strong beginning to understanding the role of the various parts of the brain in processing sensory input, and in making the body perform mechanical activity, access memory, generate creative thoughts and engage in abstract thinking in general. Abstract thought takes one of two forms: coordinated brain activity, and connectivity of thinking through logical progression. Creative thought arises from the association of ideas, the logical progression of ideas, changed views of concepts (a familiar word or idea that is evoked but viewed differently from its previous view in the mind), and serendipitous "mistakes."[36]

---

[35] Cf. the papers and books of Professor Steven Grossberg of Boston University and collaborators, and colleagues in the field.

[36] An interesting area that requires investigation is the role of writing as a memory aid. Before writing was developed, and up to the past few centuries, the human mind carried the burden of remembering a train of thought. Since writing became common there has been a remarkable increase in theoretical knowledge – particularly in the sciences – due to the interaction between thoughts in the mind and thoughts written on paper. A poignant case in point is that of Professor

Based on this line of reasoning the phrase "Mind-Body Problem" is a misnomer at best. It would better described as the Mind-Reality Problem. We create theories with our minds that describe natural phenomena. Mathematical theories of physics have been extremely accurate in describing natural phenomena – physical Reality. On a number of occasions physicists have raised the question: how can our somewhat strange mathematical theories accurately describe (or perhaps better said – accurately mirror) physical Reality. Does Reality do the computations in real wall clock time (actually more seemingly instantaneously) to make events happen according to our calculations? That is a fundamental question which metaphysics should address.

Epistemology frames our view of the knowledge of Reality and addresses the question of the nature of the theories that organize our knowledge of Reality. Thus it provides the backdrop for the study of metaphysics.[37]

2.7 The Meaning of the Words Expressing Knowledge

It is sometimes thought in Epistemology that the nature of knowledge can be reduced to words. Epistemologists define words with words and then contend with conflicting interpretations of the meaning of words.

Generally verbal descriptions do not capture the essence of the reality of physical knowledge. We have learned in Physics and, more so in Logic and Mathematics, that physical knowledge ultimately must be reduced to mathematics. The mathematical-deductive system that results must be defined with primitive terms – beyond which no explanation is possible. The knowledge embodied within the mathematical-deductive system describing an area of physics is embodied in the form of the axioms or postulates and the set of theorems/facts that follow from them. The primitive terms can either be viewed as undefined or identified as a set of specific physical entities.

From this viewpoint, discussions of the interpretation of the words describing physical knowledge are generally fruitless.

Stephen Hawking who has remarked that he regrets the impact on his work of not being able to work out thoughts on paper or a blackboard.

[37] The human mind is capable of simulating sensory input. Hypnotists do this when they cause a subject to think they see or feel something illusory. An individual can do this as well. It commonly happens in dreams. It can also happen through a form of self-hypnosis. We are familiar with this phenomena at the level of the senses. But it is possible that a sufficiently powerful mind can create a mental image of another universe and cause this universe to evolve according to some set of physical laws concocted by the mind. Then we would have a mental implementation of a form of the "many worlds" hypothesis that appears in metaphysics and modal logic. We might call this type of thought process creating "Universes of the Mind."

3. What is Metaphysics?

3.1 Definitions

Metaphysics has many definitions. Some say it is the "Theory of Ultimate Reality."[38] Others say it is the "Theory of the Nature of Abstract Entities."[39] Yet others express alternate views.

We shall take Metaphysics to be the study of the contemporary theory of ultimate Reality realizing that our views of Reality have changed drastically in the past hundred odd years and are likely to change again in the future. Rather than wait for a final "ultimate Reality" to surface[40] we will construct the Metaphysics of presently known Reality. In doing so we shall also explore the abstract entities that make up the world[41] as we know it focusing particularly on the revelations of Reality obtained in elementary particle physics which we can call Quantum Reality because its essential nature is determined by quantum physics. Quantum Reality is different from everyday Reality – the reality of everyday life with which we are familiar – that is the primary focus of most metaphysical discussions. The theory of Relativity also alters our view of Reality in a dramatic way.

We view Metaphysics as composed of three parts (See Fig. 3.1.): the Metaphysics of everyday experience,[42] Quantum Metaphysics,[43] and Cosmic Metaphysics.[44]

The combination of the mathematics of quantum theory and relativity opens a new world of metaphysics that we will call *Relativistic Quantum Metaphysics*. This new metaphysics is much simpler than traditional discussions of everyday metaphysics. Since relativity and quantum theory are of a deeper level compared to everyday phenomena we can say with assurance that Relativistic Quantum Metaphysics is closer (actually as close as we can currently get) to ultimate Reality.

---

[38] van Inwagen (2009).

[39] Lowe (2002).

[40] If perhaps it has not as yet surfaced.

[41] The word world represents our universe or the set of all universes if there is more than one.

[42] This is the metaphysics of the past 2500 years.

[43] The metaphysics of the ultra-small and of high energies whose empirical basis was only discovered in the twentieth century. Its nature is very different from "everyday" metaphysics as the following chapters will show.

[44] The metaphysics of the cosmos based on the Special and General Theories of Relativity as well as Quantum theory. This metaphysics also is vastly different from "everyday" metaphysics.

"Everyday" Metaphysics has traditionally been applied to many areas of experience including physical phenomena, thought, the mind, ethics, and Theology. Excepting Theology we view all aspects of "Everyday" Metaphysics as ultimately based on Relativistic Quantum Metaphysics. The chain connecting Relativistic Quantum Metaphysics to the various aspects of "Everyday" Metaphysics may be quite long and involved—the links of the chain in many cases are not as yet forged. Yet we will take a solidly empiricist position[45] that this chain is fact and that progress on all fronts is being made. "Everyday" metaphysics is derivative from the much deeper Reality of Relativistic Quantum Metaphysics.

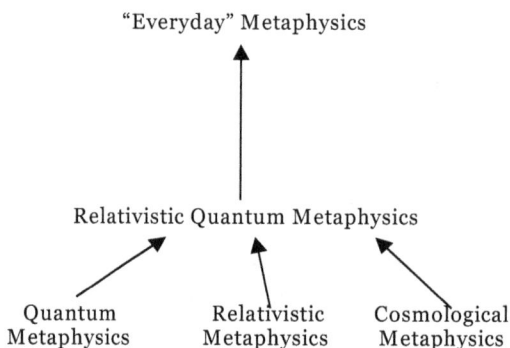

"Everyday" Metaphysics

↑

Relativistic Quantum Metaphysics

↗   ↑   ↖

Quantum   Relativistic   Cosmological
Metaphysics   Metaphysics   Metaphysics

Figure 3.1. The three sources of metaphysics and their interrelations with "Everyday" Metaphysics.

3.2 A Summary of Classical ("Everyday") Metaphysics

Classical Metaphysics[46] is the metaphysics based on our experience of everyday phenomena. Therefore it addresses issues relevant at that level of Reality.

Classical Reality contains many different substances with many different properties. As a result Classical Metaphysics is in part concerned with categorizing the many everyday substances, and exploring the interrelationships between them and their properties. It considers substratum and bundle theories relating groups of properties and substances or entities. It also considers the nature and role of propositions (essentially statements), states of affairs (situations), events, and facts.

Classical Metaphysics has found it necessary to introduce the concept of a plurality of possible worlds.[47] The "existence" of this set of

---

[45] In this choice we see our view as similar to that of Hume and other empiricists.

[46] It seems appropriate to call "everyday" metaphysics classical based on an analogy with the nomenclature of classical physics and quantum physics.

worlds enables the Modal concepts of necessity and possibility to have meaning. Is a property necessary to an entity or possibly part of an entity? – A modal question. With a many worlds scenario we can (in a simplified way) consider whether an entity in the many worlds always has a property (implying it is necessary) or sometimes has a property (implying it is contingent).

Another topic of interest in Classical Metaphysics is Causation: the study of cause and effect, their connection, and whether causes necessitate their effects. The study of classical causation is interesting because it attempts to describe these features in general terms that to this author does not reflect what actually happens in ultimate Reality according to scientific empirical data in the quantum and relativistic regimes.

Actions and events are viewed as types of entities in Classical Metaphysics. Their nature is analyzed in a qualitative, abstract manner in contrast to what we encounter at the deeper levels of actual physical Reality.

A similar statement can be made about studies of the Nature of Time and Space, as well as some studies of space-time that seek to bring in a measure of the reality of the Special Theory of Relativity. Attempts are made to abstract the nature of space and time that, in this author's view, also do not correspond to true reality. As a statement becomes more abstract, it loses content. These studies of space and time simply do not elucidate their nature.

Change takes place in time. The Classical Metaphysics of change is perhaps meaningful at the level of everyday change but does not correspond to what we have learned of change at the quantum level – the ultimate source of change.

We conclude Classical Metaphysics is limited to the subject area from which it originated – everyday objects, properties, and change. It does not transcend that area of Reality. It does not address the profound issues that Quantum Theory and the Theory of Relativity have raised. Therefore Classical Metaphysics does not address ultimate physical Reality – its stated objective. In fact, it tends to mask it with a veneer of depth and generality that it truly does not have.

---

[47] The multiple worlds concept has no relation to many universes theories in physics since the multiple worlds exist at the conceptual level and are used to clarify the notions of modality.

3.3 Are There Any Truths in Classical Metaphysics?

Van Inwagen (2009) raised two questions that support the somewhat negative tone of the previous section:[48] "Why is there no such thing as metaphysical information? Why has the study of metaphysics yielded no established facts?" To be fair there are certain principles of metaphysics that were proposed by Leibniz[49] that can be viewed as metaphysical information. Leibniz proposed three basic principles:[50]

1. The Principle of Sufficient Reason
2. The Principle of Identity
3. The Principle of Perfection

These principles, and their consequences, are to be understood in Leibniz's view as axioms within the framework of generalized subject-predicate logic. Leibniz viewed propositions (statements) about substances or entities (existents) as having a subject-predicate form[51] or as reducible to a combination of sub-statements having subject-predicate form. He used a form of analysis (*Leibnizean analysis*) of propositions that first determined the set of properties of a subject (substance) and then scanned the list to see if the predicate (a property) appeared in the list. Thus a proposition is true if the predicate is in the list of properties of a subject and false otherwise.

The Principle of Sufficient Reason is that every true proposition is analytic. Thus it asserts every true proposition can be reduced to subject-predicate form or a combination thereof, and each predicate can be shown to be in the set of the subject's properties using Leibnitzean analysis.

The Principle of Sufficient Reason is illustrative of the abstract nature of the above three principles. Leibniz's four subsidiary principles are similarly abstract:[52]

1. The Identity of Indiscernibles
2. The Principle of Plenitude
3. The Law of Continuity
4. The Principle of Harmony

---

[48] van Inwagen (2009) p. 11.
[49] Rescher (1967).
[50] Rescher (1967) chapter 2.
[51] Example: "The house is green." has house as the subject and "is green" as the predicate.
[52] See Rescher (1967) chapter 4.

The reader is directed to Rescher (1967) and other sources for information on these principles since they will not be germane to Relativistic Quantum Metaphysics. They are too general to be of value in the study of ultimate physical Reality. However Leibniz's focus on subject-predicate Logic as the underlying framework of his metaphysics will be seen to be a precursor to part of our approach to Relativistic Quantum Metaphysics.

In addition to these principles Leibniz suggested that there are many possible worlds in a physical, tangible sense – not as a conceptual construct. And he gave a criterion for the selection of our world as the world of Reality:

> The world which is most perfect … is the simplest in its axioms and the richest in phenomena.[53]

This principle is an extremal principle in the sense of extremal principles in mathematics and physics. Leibniz is famous for developing calculus (at the same time as Newton) and for his extremal principles which have been of great value in physics.

We will discuss the possibility of a selection principle for physical Reality with our universe as the solution (or a possible solution) at a subsequent point in our presentation.

We conclude our brief view of Classical Metaphysics with a figure, Fig. 3.2 that indicates where it applies in the range of Reality in this universe (world).

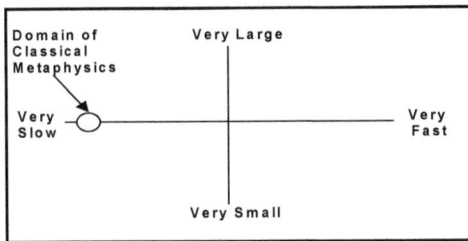

Figure 3.2. A visualization of the domain of applicability of Classical Metaphysics.

---

[53] Quoted in Rescher (1967) p. 19.

4. General Facts of Modern Theoretical Particle Physics and General Relativity

4.1 What is Physical Reality?

In the past 2500 years mankind has formulated many theories of the nature of ultimate physical Reality. Nature has constantly caused old theories to be abandoned in favor of new theories as experiments demonstrated old theories were not in accord with Reality. The theories that we have developed over the centuries – logical as they always appear to be in the light of "current" data – have continually turned out to be incorrect. Nature continues to surprise.

This situation could not be more apparent than in the major revolutions (Quantum and Relativistic) in Physics in the 20ᵗʰ century that have rocked the foundations of our previous views of Reality.

For hundreds of years Newton's theory of motion had survived unchanged. If someone in that period had pointed out that Newton's transformation between reference frames in relative motion with respect to each other was a special case of the Lorentz transformation (Special Relativity), the general reaction would have been vast disinterest since there was no experimental means to test the dynamics of relativistic velocities. Neither was there any theoretical motivation. And one could also argue that there were many possible transformation laws that approximated Newton's transformation at low velocities.[54] Thus the times were not ripe for the appearance of the Theory of Special Relativity.

Similarly Quantum theory was not in any way envisioned prior to its beginnings with the Rutherford model of the atom in 1913. Quantum theory was developed tortuously in the following decade and a half culminating in the work of Dirac. Subsequently Quantum theory has been under repeated scrutiny over the years due to its somewhat unusual implications. Nevertheless the successes of Quantum theory – particularly in the extremely accurate predictions of Quantum Electrodynamics – the most accurate predictions ever produced by a physical theory – confirm that Quantum theory is correct no matter how strange and how different its predictions are from those suggested by every day, pre-Quantum theory. Thus again we see that the logical

---

[54] There was also a belief in that period that the speed of light was infinite.

extension of physical theories of the 19th and preceding centuries would never have led to Quantum theory. Quantum theory was forced on Physics by the theoretical analyses of new experiments.

Thus if we trace the development of physics from Aristotelian times to the present, then it is clear that, at almost all times, the major changes in physical theories were unanticipated and forced on Physics by experimental results. This was particularly true in the twentieth century.

The transitions between physical theories did evidence continuity from the "old" theory to the "new" theory. The "old" theory was an approximation to the "new" theory in the region of the "old' theory's approximate validity. Examples of this continuity are the continuity from Newtonian theory to the Theory of Relativity at low velocities, and the continuity of Quantum theory to conventional mechanics at large distances.

Quantum Theory and the Theory of Relativity were developed in the first half of the 20th century. In the second half of the 20th century a fundamental theory of physics emerged called The Standard Model (of Elementary Particles) which, when combined with the General Theory of Relativity, describes what we know of Reality.

There are major remaining issues that have surfaced experimentally in elementary particle physics and in cosmology. In elementary particle physics there are predicted particles (particularly the Higgs particles) that remain to be found, and there appears to be evidence for some other new particles and new pieces of the known interactions between particles. Despite these issues the general nature of elementary particle theory is known and we can use that information as the basis for a construction of metaphysical Reality.

In cosmology, Dark Matter and Dark Energy cloud our understanding of physics in the large. We do not know what these things are. Dark Matter may be a new form of matter. It is only known through its gravitational effects at cosmological distances. Dark Energy is needed to explain the expansion of the universe. Its origin and nature are also unknown.[55] Here again, we see the great success of the astrophysical theory of cosmology (known as The Standard Model of Cosmology) as indicating that we understand much of the Reality of the universe in the large.

---

[55] There is a very real possibility that Dark Matter may cause a change in the Standard Model of Elementary Particles. Again we note that the qualitative features of the Standard Model are not likely to change, and it is upon that base that we construct Relativistic Quantum Metaphysics.

Therefore we can use the qualitative features of known 20th century Physics to establish the basic qualitative principles of Relativistic Quantum Metaphysics.

4.2 The Known Particles of Matter and Energy and their Interactions

In this section we will simply list the known particles of matter and energy found in Nature, then list known particle interactions, and lastly show some examples of interacting particles.

4.2.1 Particles of Matter

Particles called fermions constitute the fundamental building blocks of what we normally call matter. All matter is ultimately composed of a set of fundamental fermions. The fundamental fermions are spin ½ particles meaning they can be crudely viewed as "spheres" spinning at high speed. The fundamental fermion family has two subfamilies: the quarks and the leptons.

The major distinguishing feature of quarks vs. leptons is that quarks can experience the strong force (described below) but leptons cannot. The lepton family includes electrons, muons (μ) and neutrinos (v). The members of the fermion family are listed in Table 4.1.

The Quark and Lepton Families

Generation	Flavor	Quarks			Leptons	
I	1	up	u_1 u_2 u_3		v_e	electron neutrino
	2	down	d_1 d_2 d_3		e	electron
II	3	charmed	c_1 c_2 c_3		v_μ	muon neutrino
	4	strange	s_1 s_2 s_3		μ	muon
III	5	top	t_1 t_2 t_3		v_τ	tau neutrino
	6	bottom	b_1 b_2 b_3		τ	tau

Table 4.1. The known members of the family of fundamental fermions that are the building blocks of all known matter. There is some preliminary evidence for a fourth generation of fermions.

Notice also the fermions appear in three sets or generations labeled I, II, and III. There is no obvious reason for this repetition of generations.[56] We do not know at present why Nature has three generations rather than one generation. (There is some experimental evidence for a fourth generation.) It is rather like having three sets of

---

[56] Blaha (2010c) gives reasons for believing there are four generations.

china when one set would do. Each generation contains 6 quarks and two leptons.

Also notice that each quark comes in three colors. The colors are labeled with the subscripts 1, 2 and 3. For example the "down" quark actually comes in a triplet – three varieties – that we have denoted d_1, d_2, and d_3 in Table 4.1. These triplets are called color triplets. The variety of quark colors is tied up with the fact that they experience the Strong force.

Originally the three varieties of each quark type were labeled with colors such as red, white and blue. (A "red" quark was not actually red. The name only served to distinguish between the quark varieties.) Now we use simple numeric subscripts. These labels reflect internal quantum numbers that are called color quantum numbers by physicists. Consequently, the theory of the Strong Interactions of the quarks is often called Quantum Chromodynamics since it involves the "color" quantum numbers.

The three sets of color triplets are distinguished from each other by a quantum number called the flavor quantum number. Each triplet of quarks, and its corresponding lepton, has the same flavor quantum number.

The classification of quarks and leptons in the Table 4.1 is the result of over fifty years of experimental and theoretical analysis. It is like the periodic table of the elements that is so important for chemistry.

Lastly, we note that each fermion has an antiparticle with opposite charge. A fermion and its antiparticle can annihilate to produce energy in the form of photons – particles of light – or other kinds of particles.

The reader should note the remarkable simplicity of the fermion family compared to the over 100 elements in the periodic table of chemistry. *As we approach the fundamental level of Reality Nature becomes simpler!*

4.2.2 Particle Properties – Internal Quantum Numbers

The elementary particles listed in Table 4.1 are separated into two families, leptons and quarks. Each particle has distinctive features consisting of their mass and their quantum numbers. A quantum number is a positive or negative integer or simple fraction like $\frac{1}{2}$, $\frac{1}{3}$, or $\frac{2}{3}$. The simplicity of their numerical values masks our lack of knowledge of the reason for these numbers.[57] They have been found in experiments

---

[57] One can say that they are determined by symmetry groups such as SU(3) and SU(2)⊗U(1). But that then leads one to ask why these symmetry groups are relevant. The only fundamental reasons are proposed in Blaha (2010c).

starting in 1947. They are properties of elementary particles and mathematically specify the form of particles.

Originally we were familiar with mass, electric charge and spin. In the 1950's physicists realized that certain reactions between particles in experiments were allowed and certain reactions were not allowed. They then hypothesized that particles had an internal quantum number which they called strangeness and created a law that the total strangeness of the incoming particles in a reaction must equal the total strangeness of the outgoing particles in a strong interaction reaction. This type of law is called a *conservation law*.

For example, consider the following particle reaction:

$$A + B \rightarrow C + D$$

where A and B are the incoming particles, and C and D are the outgoing particles. If the strangeness of A is 1 and the strangeness of B is –1 then the total incoming strangeness is zero. Consequently the total strangeness of the outgoing particles C and D must add up to zero also. So C could have strangeness 2 and D could have strangeness –2 in this hypothetical example.

Other internal quantum numbers emerged from continuing experiments.[58] At present the known quark internal quantum numbers are:

Baryon number
Color SU(3) quantum numbers
Electric Charge
Weak hypercharge
Weak isospin
The Generation Number
Intrinsic Parity
Charge Conjugation Parity
Parity[59]

The known lepton internal quantum numbers are:

---

[58] If you wish to see the numeric values of these quantum numbers they are listed in Huang (1992) on p. 28 for quarks and p. 114 for leptons. Beyond the fact they are numerically simple integers and fractions, and of unknown origin, their values are not significant in our discussion of 21ˢᵗ Century Metaphysics.

[59] Not truly internal since it is a spatial property related to changing every spatial coordinate value to its negative.

Lepton number
Electric Charge
Weak hypercharge
Weak isospin
The Generation Number
Intrinsic Parity
Charge Conjugation Parity

Interested readers can read more about them in any of the fine introductory books on elementary particles.

4.2.3 Particles "Carrying" Interactions

Prior to the developments of 20th century physics, forces were thought to involve "something pushing something else" or as a "reaction to an action" ("To every action there is an equal and opposite reaction.")

Boson Fields ("Force Fields") of the Standard Model

Force	Number and Symbols of Bosons	
Electromagnetism and The Weak Interactions	1	W_0 Name: W boson
	3	W_i $i = 1,2,3$ Names: W bosons
Strong Interaction	8	G_i $i = 1,2, ..., 8$ Names: gluons

Table 4.2. The boson fields of The Standard Model.

The Standard Model embodies a very different point of view. Force is actually a transfer of momentum "carried" between fermion particles by spin 1 particles called bosons. When two particles interact ("collide") they exchange one or more bosons. These bosons carry momentum from one of the colliding particles to the other. Effectively this can be viewed as the particles exerting a force on each other since the momentum of each is different after the interaction. The classic view of force is that it changes the momentum of an object when exerted on the object. Thus we can say that boson particles are the "carriers" of force. The result is a picture of forces as implemented by particles – much simpler than the Newtonian view.

Here again we find the physical picture at the fundamental level is much simpler than the ruminations of classical metaphysicians on causation and change.

4.2.4 The Boson Carriers of Forces

There are twelve bosons ("forces") in the conventional[60] Standard Model.[61] (See Table 4.2.) Eight of these bosons (called gluons) embody the Strong interaction – the interaction that binds quarks together to form protons, neutrons and other familiar particles. The other four fields generate the electromagnetic force and the Weak Interaction force.

Each boson has properties similar to those of fermions. The properties of a boson are a subset of

Electric Charge
Weak Isospin
Strangeness (or alternately hypercharge)
Color SU(3) quantum numbers
Intrinsic Parity
Charge Conjugation Parity[62]

Nature appears to have four fundamental interactions or forces. In order of increasing strength they are gravitation[63] (carried by gravitons), the Weak Interactions (carried by 3 "intermediate vector bosons"), electromagnetism (carried by photons – light quanta), and the Strong Interactions (carried by the eight gluons).

4.2.5 Transitions Between Different Kinds of Particles

It is a remarkable fact that the elementary particles found it nature can transform into each other through suitable reactions. The transformation of particles into other types of particles requires both the Theory of Special Relativity and Quantum Theory in the sense that if either theory were absent particle transformations would not be possible in our universe.[64] Thus we have strong evidence that both theories are necessary for our Reality to exist as we experience it at the most fundamental level.

---

[60] Blaha (2010c) introduces a 13th Weak Interaction boson together with eight new fermions called WIMPs (Weakly Interacting Massive Particles) that may constitute the Dark Matter found in cosmology. Blaha finds these particles in the derivation of an extended Standard Model. He also derives the conventional Standard Model from a different set of postulates.

[61] These bosons are called gauge fields because they implement a local gauge symmetry. The interested reader is referred to Huang (1992), Kaku(1993) or any of the many other fine books discussing gauge theories.

[62] Only neutral bosons can have charge conjugation parity. The operation of charge conjugation transforms a particle into its antiparticle. The neutral bosons such as the photon are their own antiparticle and thus have a charge conjugation parity value.

[63] Part of General Relativity and not part of the Standard Model.

[64] This fact does not exclude the possibility of other theories enabling particle transformations in other universes.

The completeness of the set of transitions between the various types of fermions in The Standard Model is of great significance for 21st Century Metaphysics. It is evidence that all particles can be viewed as composed of one substance although each particle has a different form and, as a result, different properties. We will discuss this in detail in chapter 6. We now briefly examine examples of transformations between particle types due to its importance in demonstrating that all matter is composed of one substance.

4.2.5.1 Fermion Transitions

The fermion transitions are:

1. The set of color transitions amongst the quarks of each generation is due to the Strong interaction. Quarks exchange gluons and transform from one color to another. We can symbolize this with the expression $q_i \leftrightarrow q_j$ where q represents one of the quark types u, d, s, c, b, or t.[65] Thus $u_i \leftrightarrow u_j$ for i = 1, 2, 3 and j = 1, 2, 3 (with i not equal to j) is one of the 6 possible transitions in color of u-type quarks.
2. Quarks can also transition between types through the Weak Interaction. Thus $q_i \leftrightarrow p_i$ transitions can take place where q and p are any of u, d, s, c, b, or t. For example $s_i \leftrightarrow d_i$ transitions are allowed. There are 3 transitions of this type – one for each color. These transitions through the Weak Interaction are accompanied by the presence and participation of a pair of leptons or other quarks.
3. Leptons can transition between different types as well via the Weak Interactions. Examples are $e \leftrightarrow \mu$, $\mu \leftrightarrow \tau$, and $e \leftrightarrow \tau$. Neutrinos participate in these transitions. The three types of neutrinos can also transition between each other in an analogous manner accompanied by participating charged leptons.
4. The preceding four types of transitions also apply to the antiparticles of quarks and leptons.

We can summarize the above cases by saying any fermion or antifermion can transition to any other fermion or antifermion with possibly the participation of other particles if sufficient energy is available.

---

[65] The arrow $\leftrightarrow$ indicates that the transition or its reverse are possible under the right circumstances.

4.2.5.2 Boson Transitions

Transitions between boson types also happen.

1. The eight gluons can transform into each other through the exchange of gluons and/or with the participation of other gluons or quarks. These transformations are a result of the Strong Interaction. We symbolize these transformations by $G_i \leftrightarrow G_j$ using the notation of Table 4.2.
2. Vector bosons (W_0 and W_i for $i = 1, 2,$ or 3 in the notation of Table 4.2) can also transform from one type to another through the Electromagnetic and Weak Interactions. These transformations can take place through the exchange of vector bosons and/or with the participation of other vector bosons or fermions.
3. Transformations between gluons, photons and vector bosons with the participation of other particles are possible through combinations of the Strong Interaction, and the Electromagnetic`and Weak Interactions.

4.2.5.3 Boson-Fermion Transitions

Transformations between a boson and a fermion can take place with the participation of other particles through combinations of the Strong Interaction, and the Electromagnetic and Weak Interactions.

Thus we conclude that a transition between any two particles can take place possibly with the participation of other particles through combinations of the Strong Interaction, and the Electromagnetic`and Weak Interactions. *This combination of possibilities implies that the substance of all particles is the same (based on Ockham's Razor). Particles differ only in their form.*

We explore this conclusion further in chapter 6.

4.2.6 Some Examples of Particle Interactions

The Standard Model incorporates all of our current confirmed experimental knowledge of particles and interactions (except gravity) in a single theory.

The Standard Model consists of a complex mathematical expression called a *lagrangian* together with a procedure for quantization, and a procedure for the calculation of physical quantities. The quantization and calculation procedures are specified by Quantum Field Theory. The details of these procedures are important but they are not relevant to the theme of this book.

The lagrangian for The Standard Model can be divided into two parts. One part describes the behavior of non-interacting particles. These particles are called *free particles.*[66] After some mathematical development we can think of free particles as moving through space, and interacting with other free particles through the exchange of bosons.

As we saw bosons are the particles that carry the electromagnetic, weak, and strong forces. The photon is the boson for Quantum Electrodynamics (electromagnetism). Three W bosons are the bosons for the Weak Interaction. (The photon and the W particles are intertwined in the Electroweak theory.) Eight gluons are the bosons for Quantum Chromodynamics (the Strong Interaction).

The second part of the lagrangian describes the interactions of particles. These interactions typically involve the exchange of bosons between interacting fermions (and bosons). The interactions on the quantum scale become the forces we see in everyday life when vast numbers of quantum interactions take place.

Electron Emits Photon   Electron Absorbs Photon

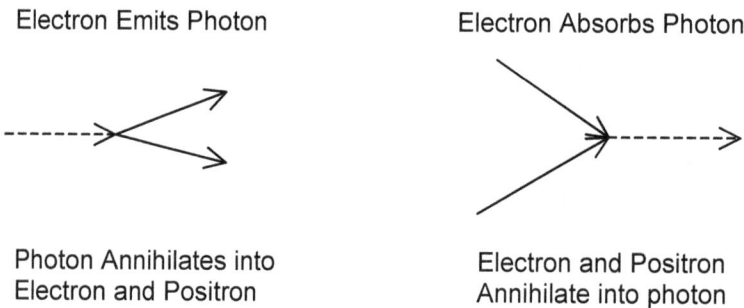

Photon Annihilates into   Electron and Positron
Electron and Positron   Annihilate into photon

Figure 4.1. Some Simple Electromagnetic Interactions. Solid lines are electrons and positrons. Dashed lines are photons – quanta of light.

It is an interesting point of history to note that hundreds of years ago scientists thought that light was made of particles and that seeing involved the reception of particles. Forces were also viewed as the result of myriads of particles acting on a body. These ideas were discarded in the nineteenth century only to be revived in the latter part of the twentieth century in a much deeper and more detailed quantitative form.

The interaction part of The Standard Model specifies many interactions between particles. We will look at a small representative sample of these interactions to obtain an understanding of their basic

---

[66] Quarks appear to be always bound inside protons, neutrons and so on. However under the right experimental conditions they can be seen as quasi-free for very short periods of time.

idea and then use the concept these interactions in our development of 21ˢᵗ Century Metaphysics.

One of the simplest examples of particle transformations is the transformation of a photon into an electron- positron pair.[67]

The interactions depicted in Fig. 4.1 correspond to an amazing feature of matter: particle creation and annihilation with the conversion of energy (photons) into matter and vice versa. As noted earlier this energy-matter conversion requires both the quantum theory and the theory of Special Relativity to work. Special Relativity is required because of the conversion between matter and energy. The creation and annihilation process is an inherently quantum process requiring Quantum Field theory.

The interaction processes in Fig. 4.1 are among the simplest electromagnetic interactions of an electron. These simple interactions can be combined to make an infinity of more complex composite interactions. For example two electrons can scatter off each other by exchanging a photon. The photon exchange takes place by having one electron emit a photon and the other electron absorb it. (See Fig. 4.2.)

Another example resulting from the combination of these simple electromagnetic interactions is the case of two incoming electrons scattering (colliding) and producing three electrons and one positron (the output particles). (See Fig. 4.3)

Electrons can interact by exchanging photons. A pair of generated photons can combine to generate an outgoing electron and positron pair.

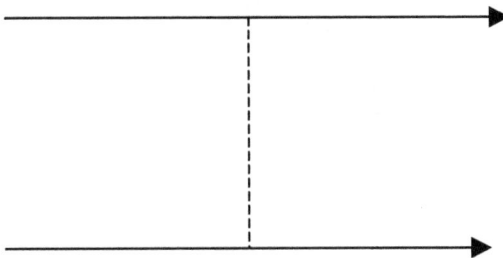

Figure 4.2. A diagram for two electrons interacting by exchanging a photon (dashed line).

These examples have all been based on the electromagnetic interaction. We now consider an example of a Weak Interaction process

---

[67] A positron is the antiparticle of an electron.

in which a muon μ decays (transforms) into an electron e plus a muon type neutrino ν_μ and an electron type antineutrino ν_e. The decay is diagrammed in Fig. 4.4.

Another example of the Weak interaction implementing a transition between particle types is the transition (decay) of a neutron into a proton plus other particles (leptons).

A neutron consists of a u type quark plus two d type quarks all bound together by the Strong Interaction (the gluons embodying the Strong Interaction are not displayed in Fig. 4.5). A d quark transforms into a u quark and other particles (leptons). The bound group of three quarks, a d quark and two u quarks, constitutes a proton.

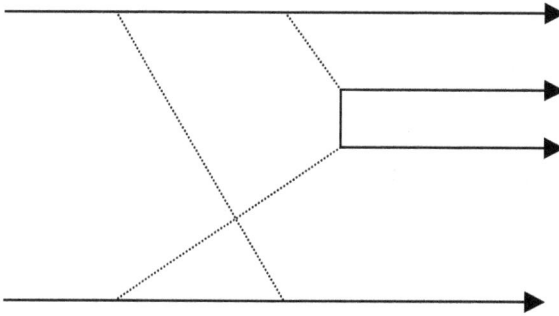

Figure 4.3. Two electrons collide producing 3 electrons and a positron. The dotted lines are photons.

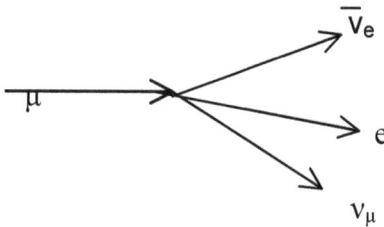

Figure 4.4. The decay (transition) of a muon into an electron plus other particles (neutrinos).

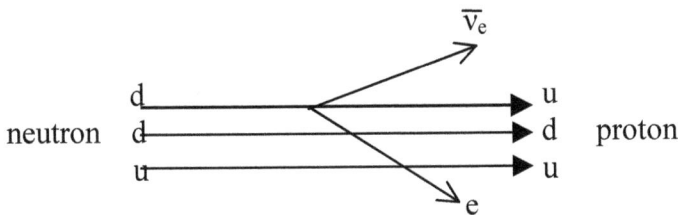

Figure 4.5. The decay (transition) of a neutron into a proton plus an electron and an electron type antineutrino. One of the d quarks in the neutron decays to a u quark with the emission of leptons.

These examples show how interactions in The Standard Model can produce a complex variety of transitions between particle types.

4.2.7 Localization of Particle Properties by "Gauging" Bosons

Yang and Mills[68] observed that the internal quantum number properties[69] of particles are local properties in the sense that they can have different values at different space-time points. For example, the charge of a particle can be different at different space-time points. This feature of particle properties, which at first glance seems strange, is made possible by "changing" the interactions at different space-time points in sync with the change in internal quantum numbers. The simultaneous, synchronized change of particle properties and particle interactions at different space-time points causes the net experimental results to be the same as they would have been if the properties were not changed.

The changes in the interactions are implemented by a theoretical physics process called changing the "gauge" of the spin 1 bosons that carry the interactions.[70] Consequently the bosons in Table 4.2 are called gauge bosons because they have a changeable gauge in The Standard Model.[71]

---

[68] C. N. Yang and R. L. Mills, Phys. Rev. **96**, 191 (1954).

[69] Actually they only considered charge and isotopic spin since the understanding of other internal quantum numbers was still unclear in 1954.

[70] The gauge features of The Standard Model are difficult to describe without resorting to fairly complex mathematics. The reader who wishes to learn more on this topic is referred to Kaku (1993), Huang (1992) and other texts on the Standard Model.

[71] The Weak Interactions have features which "break" the gauge feature of particle theory and so, due to this Weak Interaction feature, one cannot arbitarily change the internal quantum numbers to which the Weak Interaction is sensitive at differing space-time points.

Particle internal quantum number properties, in general, depend on the space-time "location" of the particle. We call this space-time dependence *locality*. The locality of particle properties will be significant when we discuss substances and their properties in chapter 6, and also when we describe a derivation of The Standard Model from Logic subsequently.

4.2.8 General Relativity – Gravitation

The General Theory of Relativity developed by Einstein provides a satisfactory description of the universe in the large, and has passed significant experimental tests over the past ninety years. It is not a quantum theory and attempts to create a quantum theory of General Relativity have encountered technical difficulties. Thus uniting it with The Standard Model is at present problematic.[72]

However research efforts have suggested that a successful quantum General Relativity will eventually be developed. One aspect of a quantum General Relativity theory on which most physicists agree is that there will be a boson particle, the graviton, that will be the carrier of the gravitational force in a manner analogous to the carrier bosons of the Weak Interaction force, the Electromagnetic Interaction force and the Strong Interaction force.

Thus we conclude that matter, energy, and forces are all particulate in nature.

4.2.9 Space and Time

Since General Relativity deals with space and time it must in some sense determine explicitly, or implicitly, what space and time really are.

Mathematically one can consider space as having coordinates that are specified by using a standard ruler to measure distances. Similarly time is measured by using a standard clock that measures time intervals.

But the questions remains what are space and time and how do they differ. The Theory of General Relativity provides an answer to these questions. The answer seems to be that space and time are similar in nature with one exception. We call time the dimension in which all entities resident in the universe increase their coordinate value for that dimension, or all entities decrease in their coordinate value for that dimension. More concretely, everything "ages." A dimension that we call a space dimension has the feature that an entity in the universe can move backwards or forwards in that dimension.

---

[72] This author has developed a form of Quantum Field Theory that enables a successful quantum theory of gravitation to be formulated. See Blaha (2005a) and (2010c).

The Schwarzschild solution of General Relativity for massive bodies such as black holes illustrates the view that space and time dimensions are similar in nature. Inside the black hole (when the radius r is less than 2MG[73]) the time coordinate becomes like a space coordinate and the radial coordinate r becomes like a time coordinate. When particles enter the black hole their radial coordinates r decrease in value until the particles reach the black hole's center at which the black hole singularity resides.

Thus we conclude that space and time are the same except for that one difference.[74] This conclusion is further supported by the Special Theory of Relativity where "rotations" can take place between space and time – again confirming that space and time are of the same type except for the one aforementioned difference of property.

Consequently, physicists refer to space-time realizing the unity of their basic nature.

As to the question of what is space-time, contemporary physics provides an answer that seems to accord with Reality.

First consider a universe (not terribly dissimilar to our own) governed by The Standard Model and General Relativity with perhaps isolated clumps of matter (planets, stars, galaxies, black holes and so on). This universe has much "empty space" that appears to contain nothing. However the "empty space" really constitutes the vacuum of The Standard Model and thus has two important features: 1) at every point in space there are an infinite number of negative energy fermions of each of the quark and lepton particle types,[75] 2) at every point in space quantum fluctuations occur in a random way generating particle-antiparticle pairs that exist for an instant and then recombine (and "disappear").

Normal particles and their aggregates – the masses in the universe – have positive energy and are randomly scattered throughout the universe. These masses distort the space-time around them, and, in total, cause the universe to curve in such a way as to be a closed universe.

Thus the universe is a curved, closed, quantum surface or manifold. Space-time then can be thought of as the coordinates

---

[73] M is the mass of the black hole and G is the Newtonian gravitational constant.

[74] There is much discussion of the "arrow of time" in the popular and scientific literature. The "arrow of time" in the black hole example is set by the location of a singularity in the space-time coordinates at the center of the black hole. The "time" decreases. In our universe, outside of black holes, the "arrow of time" is opposite to that in black holes. Time increases.

[75] Consequently each point of "empty space" has an infinite density of particles. Nothingness does not exist in our universe!

(numbers) labeling the points on this manifold rather like latitude and longitude label locations on the earth's surface.

5. Some Qualitative Principles of The Standard Model and General Relativity

Chapter 4 provided a qualitative description of the features of The Standard Model and Gravitation that omits mathematical details in the interest of a succinct description that would be readable by the philosopher community and by the general public. In this chapter we abstract the "philosophic/metaphysic" physics content of chapter 4 in the form of a set of features of ultimate Reality that we will use in chapter 6 to define a 21ˢᵗ Century Natural Philosophy.

The set of Physics properties that establish the basis of our new Natural Philosophy are:

1. Matter and energy are both composed of particles but not the hard little spheres of everyday imagined particles.

2. Matter can be converted into energy and vis versa due to the combined effect of Quantum Theory and the Special Theory of Relativity.

3. All particles can undergo transformations between any of the various different types given sufficient energy and with the possible participation of other particles. Thus particle A plus other particles can transform into particle B plus other particles for any particles A and B listed in Tables 4.1 and 4.2.

4. Based on property 3 we can assume all particles are made of the same substance. This substance can be created in any quantity given sufficient energy. This substance has an intangible nature but in the form of particles becomes the matter and energy that we observe experimentally.

5. Particles interact via the exchange of particles. This is the essence of force.[76]

---

[76] It "explains" change and causality in metaphysics.

6. There are certain conserved quantities (internal quantum numbers) such as charge that must be conserved in any interaction. There are other internal quantum numbers that are almost conserved except for the effects of the Weak Interactions. These quantum numbers serve to give elementary particles stability against transformations. They also distinguish between the various types of particles.

7. The known elementary particles that compose matter are the spin ½ fermions listed in Table 4.1. The set of their properties are internal quantum numbers (section 4.2.2), mass, and spin.

8. The known elementary particles that have spin 1 are called bosons and are listed in Table 4.2. Photons are usually what we call quanta of energy. They are the carriers of the electromagnetic force. Intermediate vector bosons are the carriers of the Weak Interaction Force. The eight gluons are the carriers of the Strong Interaction force.

9. The behavior of elementary particles (except gravitons) is described by The Standard Model which is a Quantum Field Theory.

10. A Quantum Field Theory specifies the interactions between particles. It also describes the quantum fluctuations in which particles are created and destroyed in particle interactions.

11. The vacuum is composed of the same intangible substance as particles but without any of the space-time or internal quantum numbers that specify the form of particles. Particle-antiparticle pairs constantly are created and destroyed in the vacuum as quantum fluctuations. The Standard Model implies the vacuum is a very dynamic entity filled with infinite numbers of particles, that we can only detect indirectly, that are continuously undergoing quantum fluctuations.[77] The vacuum is filled with a sea of the various kinds of fermions. Because the sea of fermions is uniform it is not directly detectable.[78]

---

[77] The Casimir effect is one experimentally measurable effect of the structure of the vacuum. It is an electromagnetic phenomenon that occurs when two uncharged metal plates are placed a few micrometers apart in a vacuum with no external electromagnetic field present. The presence of the plates changes the structure of the vacuum causing a force to appear due to quantum effects.
[78] For example, the infinite number of electrons in the sea is uniformly spread through space and so the net force of the sea on an isolated electron not in the sea is zero because the forces exerted by all the electrons in the sea balance each other so the net force is zero.

12. Property 11 implies that our entire universe is filled with "something". There is no region in the universe that can be described as containing nothing although seemingly empty regions are colloquially described as containing "nothing."

13. The internal quantum numbers of a particle, in general, depends on the space-time "location" of the particle. This space-time dependence is called *locality*. Locality is implemented by using gauge bosons as the carriers of particle interactions. The allowed variation of internal quantum numbers from point to point in space-time is compensated by a change in the interaction from point to point in such a way that the resultant physical effects are the same as if no variation or change took place.

14. General Relativity has a spin 2 particle carrying the gravitational force called the graviton. [79]

15. Space and time are dimensions in space-time that differ only in that time always increases, and one can go to and fro in spatial dimensions. In the Special Theory of Relativity space and time can be rotated and thus "mixed" together. In the General Theory of Relativity the time dimension can become a space dimension, and a space dimension can become a time dimension.[80] The numerical values of space and time mark points in space-time just as latitude and longitude mark points on the earth's surface.

---

[79] Quantum theories of General Relativity remain the subject of much debate.
[80] This situation occurs inside Black Holes. See Blaha (2004).

6. Fundamental Concepts of 21ˢᵗ Century Metaphysics/Natural Philosophy

6.1 Reduction of Reality to the Presently Known Ultimate Reality

Classical Metaphysics was developed over the millennia based on concepts that arose in everyday experience with a small admixture from science (up to the nineteenth century.) Since 1900 there has been an explosion of knowledge – primarily scientific knowledge – that calls for a new metaphysics/Natural Philosophy of physical Reality.

We have come to realize that everyday phenomena which appear complex in nature and involve a staggering variety of different substances, properties, and events at the physical level, the mental level and the social level are all ultimately derivative[81] from a vastly simpler underlying physics that is embodied in The Standard Model of Elementary Particles and the General Theory of Relativity.[82]

Chapter 5 contains a summary of the qualitative nature of this "all encompassing" physics. Based on this summary we can consider the progressive depth of physical Reality as:

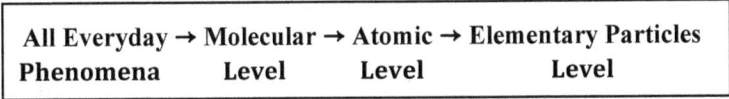

All Everyday →	**Molecular** →	**Atomic** →	**Elementary Particles**
Phenomena	**Level**	**Level**	**Level**

omitting gravity for a while from our discussion. All substances at the everyday level with all their properties are ultimately consequences of the nature and behavior of elementary particles. The set of known elementary particles, their nature and their behavior, are extremely well described by The Standard Model of Elementary Particles.

We can view the progressive depth of physical Reality from the everyday level to the level of the universe(s) as

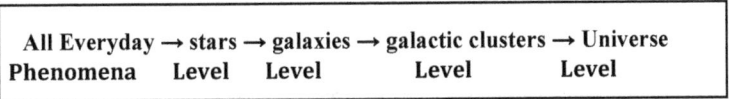

All Everyday →	**stars** →	**galaxies** →	**galactic clusters** →	**Universe**
Phenomena	**Level**	**Level**	**Level**	**Level**

---

[81] We will not consider Theological questions and the nature of God because we view this area as not part of physical Reality although the spiritual and the physical may interact with each other.

[82] We say that knowing that aspects of gravity and the cosmos (Dark Matter and Dark Energy for example) as well as features of elementary particles (including the issue of whether there is a deeper theory of elementary particles such as a Superstring theory) remain to be understood. These unknowns are not relevant to "everyday" phenomena and thus our development of Relativistic Quantum Metaphysics does provide a complete underpinning for "everyday" phenomena and the domain of Classical Metaphysics.

General Relativity has major effects at the levels of stars, galaxies, galactic clusters, and the universe.

With these views of Reality in mind we will develop a metaphysics/Natural Philosophy based on the qualitative features of The Standard Model and on the theories of Special and General Relativity.

6.2 Particles and Substance

In everyday Reality we see a multitude of substances. At the elementary particle level we know of the existence of quarks, leptons and gauge bosons.[83] They are listed in Tables 4.1 and 4.2. The question arises whether each particle is a separate substance or whether the particles are better viewed as different forms of one fundamental substance.[84]

There is good reason to believe that particles are composed of one fundamental substance, and that this substance can take a variety of forms – each form being one of the elementary particles. We come to this view because of the ability of particles to transform into each other. (Property 4 of chapter 5). Particles can transform into each other through the decay of one particle into other particles, and can transform into each other when they interact (or exert a force) on each other. Although they differ in their properties, their ability to undergo transformations into each other strongly supports the idea that there is one substance that can take thirty-seven[85] (or more) different forms – each with different properties.

Can we take the alternate view that there are thirty-seven different substances – each particle constituting a different substance? Yes, *but* Ockham's Razor – the simplest solution is usually the correct solution – would support the view that there is one substance capable of assuming thirty-seven different forms.

And so we come to the remarkable conclusion that ultimate Reality can be based on one substance and assert the first principle of 21ˢᵗ Century Natural Philosophy:[86]

---

[83] There may also be other particles as yet undiscovered such as Higgs particles or particles predicted by string theory. The considerations of this chapter would not be changed if any of these particles were found.

[84] The nature of Dark Matter and Dark Radiation is an open question. They may constitute another substance or possibly two substances. However experimental results reported in the past month (October, 2009) suggest that Dark Matter can transform or decay into normal matter and energy. If these results are correct then Dark Matter is of the same substance as normal matter.

[85] Assuming three generations of fermions.

[86] SuperString Theories, one set of possible theories of ultimate Reality, assume all elementary particles in the universe are made of extremely small mathematical strings. Thus these theories conform to principle I.

I. There is only one fundamental substance in the world (universe). It can assume a variety of forms as particles, which we call fundamental elementary particles.

This principle harks back to the monistic ideas of Thales (and his successors such as Anaximander and Anaximenes of Miletus). The substance is not water or air as the 6th century BC philosophers proposed, but does bear comparison with the *apeiron* of Anaximander in that it is best described as "infinite" and without "distinctive" features where "infinite" means creatable without limit and "distinctive" means not describable in terms of everyday experience. Following Anaximander in part we postulate:

2. The one fundamental substance is creatable without limit given sufficient energy. It is not describable except through quantum, relativistic mathematical equations. It is inherently quantum in nature and undergoes constant quantum fluctuations.

The "amount" of substance in particles can be indefinitely expanded through particle interactions such as those displayed in Figs. 4.1 – 4.5.

The various forms it can assume are the elementary particles and the vacuum. The vacuum also consists of this substance but with a different form and different properties from those of individual particles. (However the properties of the vacuum are related to the properties of particles since the vacuum consists of infinite numbers of particles at every point. We discuss the vacuum in section 6.4.)

Having reduced the number of substances to one substance, the question naturally arises what is that substance? The answer is that it is unknowable (postulate 2) except for its properties and the forms it can assume.

The properties of the substance are known but the substance itself is an unknowable except mathematically.[87] We can only specify the forms it takes and their properties. Taking a note from Logic we simply call it a *primitive term* – an undefinable term which we can use to build a mathematical theory. To the reader who might object to this answer we point to Euclid's geometry, which has several primitive terms in its axioms. Thus primitive terms are to be expected in a theory. Some day if

---

[87] One can describe it by analogy with familiar everyday substances but that exercise does not in the end lead to further insights into the nature of the fundamental substance.

a more fundamental theory than The Standard Model appears, then it also can be expected to have primitive terms as well.

6.3 Fundamental Substance: Forms and Properties

Classical Metaphysics is much concerned with properties. It considers them individually for entities, in bundles of properties that entities possess, and as forming an entity in the form of a substratum.

We have learned from the study of the physical nature of matter that the thirty-seven plus known elementary particles have bundles of properties consisting of energy, momentum, spin, and internal quantum numbers. We will call the bundle of properties of a particle its *form*. Thus we have a third principle of Natural Philosophy of great simplicity compared to the consideration of everyday properties in Classical Metaphysics:

3. Each particle's form consists of a finite number of properties: its energy, momentum, spin, and internal quantum numbers.

The internal quantum number properties of particles are, in general, *local* meaning that they can vary from space-time point to space-time point. When they vary from space-time point to space-time point, their particle interactions correspondingly change[88] so that there are no physical results of the variation. It is as if the internal quantum number properties were unchanged from space-time point to space-time point. However, there is an important effect of locality. The interactions between the particles have to be defined in such a way as to allow locality. As a result the bosons that carry particle interactions are gauge bosons – bosons whose features and interaction are specified by Yang-Mills gauge theory. Yang-Mills gauge bosons implement locality.

The extreme simplicity of the theory of particles and their properties makes the theoretical developments in the study of properties in Classical Metaphysics not relevant to the ultimate Reality of the world (universe) that we inhabit.

However, Classical Metaphysical studies may be relevant to the study of other possible worlds (universes) if they are applied in a wider framework than the picture of ultimate Reality of our universe. In that wider framework Relativity and Quantum theory, may be supplemented, or replaced, by other theories. Thus a wider metaphysics would be required and the considerations of Classical Metaphysics might be relevant. We will not consider this point further in the present work.

---

[88] Through a mathematical technique that physicists call a gauge transformation.

6.4 The Vacuum

The Standard Model has a vacuum which is composed of an infinite number of all the different particles. But these particles are uniformly distributed throughout space so the forces that they might have exerted on any individual particle of matter cancel each other[89] and thus we think of the vacuum as empty. The vacuum is also known to be filled with a sea of the various kinds of fermions. Because the sea of fermions is uniform it is not directly detectable. For example the infinite number of (negative energy) electrons in the sea is uniformly spread throughout space and so the net force of the sea on an electron (having positive energy and so not in the sea) is zero because the forces exerted by all the electrons in the sea on the isolated electron balance each other and completely cancel.

The vacuum, therefore, is also composed of the substance of particles in infinite, uniform, quantity.

At all times and points the vacuum is undergoing quantum fluctuations creating a particle-antiparticle pairs for an instant. Thus the vacuum has a dynamic aspect. But the extremely short existence of the particle-antiparticle pair makes vacuum fluctuations unobservable except indirectly in their effects. We know that they happen because they affect the observed properties of matter.

An example of a vacuum effect is the Casimir effect[90] - an experimentally measurable effect of the structure of the vacuum.

Thus our entire universe is filled with "something". There is no region in the universe that can be described as containing nothing although seemingly empty regions are colloquially described as "a vacuum containing nothing." Nothing is something.

6.5 Change, Interactions and Causation

Causation and change have been much studied in the abstract in Classical Metaphysics. The abstractness of the work in these areas and the variety of everyday phenomena has led to a plethora of ideas. But this plethora is little related to the ultimate Reality embodied in Relativistic Quantum Field theory in general and The Standard Model of Elementary Particles in particular – the deepest verified theory of physical Reality – supported by a large amount of positive experimental

---

[89] This is a simplification in that a charged particle, for example, distorts the vacuum attracting particles of opposite charge to it which "mask" (renormalize) part of its charge.

[90] The Casimir Effect is an electromagnetic phenomenon that appears when two uncharged metal plates are placed a few micrometers apart in a vacuum with no external electromagnetic field present. The presence of the plates changes the structure of the vacuum causing a force to appear due to quantum effects.

data with no significant contradictory experimental data.[91] (The Standard Model is a specific Quantum Field Theory.)

In The Standard Model causation and change result from the interactions (forces) between elementary particles. The interactions can take place at very small distances, or at large distances, between individual elementary particles. When particles combine to form large (everyday) bodies the individual particle interactions can accumulate to cause tangible effects at everyday distances between everyday objects.

At the most fundamental level the decay of an elementary particle, and the interaction between two elementary particles, take place through the exchange of the thirteen fundamental gauge bosons mentioned earlier: gravitons, photons, three vector bosons, and eight gluons. They can be visualized using Feynman diagrams. These diagrams, developed by Richard Feynman, associate a straight line with each of the twenty-four "matter type" particles and a "wiggly" line with each of the thirteen "interaction particles." They are viewed from left to right. (Time can be considered as increasing from left to right in these diagrams although that is not strictly correct.) See Figs. 6.1 – 6.3 below for simple Feynman diagrams illustrating decays and interactions.

Thus we see again that the complex issues and discussions of Classical Metaphysics are reduced in Relativistic Quantum Metaphysics to interactions between particles or combinations of particles – much simpler, less abstract, and not subject to dispute. Causation and Change are reduced to single or multiple particle interactions.

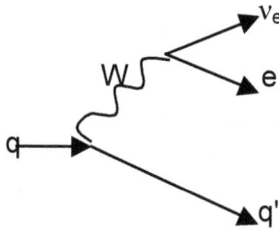

Figure 6.1. A Feynman Diagram for "Causation" and "Change" in The Standard Model. It depicts the decay of a heavy quark particle into three particles: a lighter quark q', an electron e and an electron-type neutrino v_e. It illustrates one form of Change.

---

[91] We state again that it is known that there are areas that are not part of the Standard Model and there may be a deeper theory waiting in the wings. These difficulties are not roadblocks to viewing the Standard Model provisionally as ultimate physical Reality since The Standard Model will still hold for current experimental data within a larger theory or a deeper theory. In these cases it becomes a good approximation for its domain of experimental data.

Figure 6.2. Another Feynman Diagram for "Causation" and "Change" in The Standard Model. It depicts the interaction (force exchange) between two particles which can lead them to change direction or to change the type of particles that they are. There are two simple diagrams shown. Actually these are the simplest diagrams of an infinite set of diagrams that contribute to the change in the two initial particles to the two final particles.

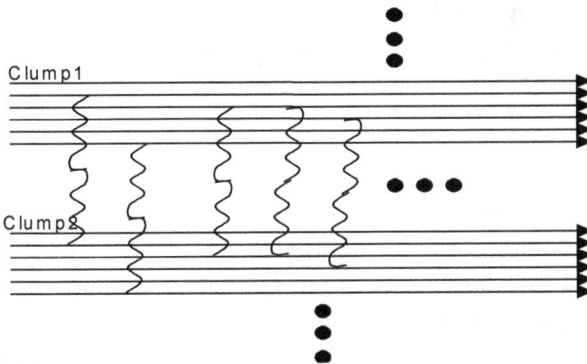

Figure 6.3. Yet another Feynman Diagram for "Causation" and "Change" in The Standard Model. It depicts the gravitational interaction (force exchange) between two clumps of matter via graviton particles on particles within the clumps, which lead the clumps to change direction. This illustrates the force of gravity at the quantum level. As yet the understanding of quantum gravity is subject to dispute. But most physicists would agree that gravitons exist and are the fundamental "carriers" of the force of gravity. Again these are the simplest diagrams of an infinite set of diagrams that contribute to the process of change. Clump1 and Clump2 each consist of a large number of particles in lumps of matter. The thick "dots" represent lots of particles and lots of gravitons being exchanged between the particles in the clumps. The result of the extremely large number of gravitons exchanged

is that the clumps change direction and approach each other since gravity is a strictly attractive force according to the General Theory of Relativity.

The general, abstract approach of Classical Metaphysics to Causation and Change is defective in not properly dealing with quantum phenomena.[92] However if one wishes to study all possible universes then Classical Metaphysics, suitably generalized to include Quantum and Cosmological Reality, may support the study of alternate universes. This would open a new sphere of activity for classical metaphysics.

Considerations of the sort illustrated by Figs. 6.1 – 6.3 lead us to the following additional principle for Relativistic Quantum Metaphysics:

4. *At the level of ultimate Reality, Causation and Change take place through interactions (forces) mediated (conveyed) by particles.*

6.6 Natural Philosophy of Space-Time

We view the metaphysics/Natural Philosophy of space-time to be adequately described by section 4.2.9 and properties 14 and 15 of chapter 5.

6.7 Modality

Modality is concerned with questions of necessity and contingency. Is a property necessary to an entity or is a property merely a possible part of an entity? Modality appears in Classical Metaphysics in relation to questions about the nature of entities.

Part of the framework of Modality is the existence of multiple worlds at least in a conceptual sense. Thus if a certain entity exists in all worlds and always has a certain property then that would imply that the property was *necessary* to the entity. If, on the other hand, a certain entity exists in all worlds and has the property in some worlds but not in others then the property would be a *contingent* part of the entity.

Questions of Modality appear when we consider Relativistic Quantum Metaphysics. There may be situations where a property may be necessary or contingently part of an entity.

When we consider the range of possible universes that exist within the conceptual framework of Physics we find that Modality in its conventional form is not applicable. *The strict division between the*

---

[92] For example Leibniz and others suggest that Change must be continuous. It is a known fact that electrons make jumps ("quantum leaps") between orbits in atoms. They are never between atomic levels. Quantum Theory shows the inadaquacy of the Classical Metaphysics views of Causation and Change.

contingent and the necessary in Modality is not relevant in Physics because a consideration of the range of possible physically acceptable universes proves that all physical laws, properties and entities are contingent.

Some examples that demonstrate that all physics is contingent are:

1. A completely empty universe devoid of matter, radiation, quantum theory and relativity theory is a physically acceptable universe in that it is internally consistent.
2. A universe consisting of one point without quantum effects is a physically valid universe.
3. A "flat" universe containing matter and/or energy with or without quantum theory and/or relativity theory is an internally consistent universe and thus physically acceptable.

These examples of possible universes show that physics is entirely contingent and thus conventional Modality is not relevant for Physics.

However there is an extension of Modality that does not seem to have been explored – a third modal category – *contingently necessary* – in addition to the categories of contingent and necessary. An entity or property is contingently necessary if it is required due to the existence of another contingent or necessary entity or property.

An example of a contingently necessary property in our universe appears in the phenomena of particle decay and transformations.[93] For these phenomena to take place in our universe both Quantum Theory and the Theory of Special Relativity must hold. Without both theories particles could not transform into one another. Thus Quantum Theory and the Theory of Special Relativity are contingently necessary for the phenomena of particle transformations to take place. If we did not have particle transformation phenomena in our universe, then Quantum Theory and the Theory of Special Relativity would not be contingently necessary.[94]

So we conclude that an extended Modality embodying the concept of contingently necessary is relevant for 21ˢᵗ century Metaphysics/Natural Philosophy.

---

[93] We used the particle transformations of The Standard Model to establish the principle that all particles are composed of the same substance. These phenomena are contingent in the sense that universes could exist in which they did not exist.

[94] Although there are other phenomena that would make them contingently necessary.

6.8 Matter and Being

In sections 6.2 and 6.3 we showed how matter and radiation are composed of a single substance which can take a variety of forms. We now wish to get to the heart of the substance of matter. What is a particle made of? It appears that a particle has form without substance (as we commonly understand substance). Its form is particulate in part and wave-like in part.[95] We might think of it as "nothingness"[96] upon which a semi-permanent (or permanent) form is imposed with the quality of "Being." Being consists of existence for some interval of time and implies observability. An entirely unobservable entity could have a persistent form, but, lacking observability, could not be considered real since there would be no manifestations of it in our Reality. Fortunately, for us Quantum Theory requires an observer, and observability, for all real entities and events. Thus all known particles interact and that makes of them one substance – our Reality.[97]

Particles exist and possess Being. Particles can interact with each other to create new particles or to annihilate into radiation.[98] So the forms imposed on nothingness (particles) can transform to other forms but do so in such a way that certain features of these forms are preserved. These features satisfy what we call conservation laws such as the conservation of energy.

It is rather remarkable that particles can undergo true creation and annihilation because these features represent transformations between *Being* and *non-Being*. Seeing the process of true creation and annihilation in the laboratory it is clear that Being is an acquirable property. As such, the Big Bang, which seems to be the origin of the universe, can be accepted as a transition from nothingness to Being through an expansion from a point to space-time differentiated regions with a variety of forms of particles. Thus we can perceive the nature of the process although the precise details remain to be determined.[99]

---

[95] A true definition of nothingness seems to be lacking. Attempts to define nothingness tend to degenerate into circularity. It appears that nothingness is best considered to be a primitive term – undefinable but identifiable with the everyday view of nothingness.

[96] Earlier we suggested that the one substance be considered a primitive term. In this section we attempt to go one step further and develop a physical/conceptual identification of the one substance.

[97] This includes Dark Matter, which interacts with normal matter through gravitation and perhaps through other forces very weakly.

[98] Radiation is composed of particles as well: photons, ElectroWeak bosons, Strong Interaction gluons, and gravitons. Gravitons have not as yet been detected due to the weakness of the force of gravity.

[99] Blaha (2004) describes a theory of the quantum Big Bang.

7. Fundamental Physics Theories

> The further we pursue these inquiries, the fewer become the
> primitive truths to which we reduce everything; and
> this simplification is inself a goal worth pursuing. ...
> The aim of proof is, in fact, not merely to place the truth of
> a proposition beyond all doubt, but also to afford us an
> insight into the dependence of truths upon one another.
> Gottlob Frege – *The Foundations of Arithmetic*

7.1 What is the Form of the Fundamental Theory of Physics?

Many individuals have speculated on the ultimate form of the fundamental theory of physics, often called the "Theory of Everything". Some say it is a form of string theory. Others have suggested a theory based on a discrete lattice-like space-time with, perhaps, extra dimensions. Yet others have suggested a self-organizing theory from a primordial chaos. And, lastly, a more conservative group has suggested theories based on quantum field theory.

In all cases the fundamental theory of physics must take the form of a set of postulates that use primitive terms that are either loosely defined or "intuitive" or both. The implications of the postulates are explored and experimentally testable consequences ultimately derived.

It is clear from the experimental success of The Standard Model of Elementary Particles (modulo a few discrepancies and some major open questions[100]) that The Standard Model must be "derivable" or "constructible", at least approximately, for currently accessible energies from THE fundamental theory when it is found.

7.2 A Theory for its Time

In the time of Charlemagne Western European scholars obtained access to the literature and philosophy of classical Greek times. A mini-Renaissance, the Carolingian Renaissance, followed that expired after a relatively short time because the intellectual climate of the times in Western Europe "was not ready for it." In contrast, at a much later time, the Italian Renaissance flowered, and was culturally diffused throughout a receptive Western Europe, beginning the modern age of culture and science. For a cultural or intellectual advance to succeed, and not wither, the intellectual/cultural climate must be in a receptive state.

---

[100] Such as how Dark Matter and Dark Energy relate to The Standard Model.

In the case of physics a similar situation holds. Consider the case of Newton's theory of motion, which survived unchanged for hundreds of years. If someone in that period had pointed out that Newton's transformation between reference frames in relative motion with respect to each other was a special case of the Lorentz transformation, the general reaction would have been vast disinterest since there was no experimental means to test the dynamics of relativistic velocities. Neither was there any theoretical motivation. And one could also argue that there were many possible transformation laws that approximated Newton's transformation at low velocities.[101] Thus the times were not ripe for the appearance of the Theory of Special Relativity.

More generally, if we trace the development of physics from Aristotelian times to the present then it is clear that, at all times, the major changes in physical theories were unanticipated and generally driven by experimental results. This was particularly true in the twentieth century. It is also true in each historical period of physics, that when a successful theory was constructed, it described experimental results as they were understood in that period. The transitions between physical theories evidenced a continuity that was reflected in the "old" theory being an approximation to the new theory in the region of the "old' theory's approximate validity.

The drift of the above considerations is clear. Having been driven to The Standard Model by over seventy years of experimental and theoretical work, it is incumbent on theoretical physics to make sense of its form and origin.

Countless negative comments about the peculiarity of its form have appeared since the general form of The Standard Model was established around 1975. Attempts to develop a more fundamental theory of which The Standard Model is an approximation, have been mostly SuperString theories. Lacking experimental guidance, and limited by the complexity and variety of the mathematics of SuperString theories, progress along these lines has been somewhat limited and not experimentally substantiated.

Thus we still face the issue of explaining The Standard Model. It certainly accounts for the vast bulk of experimental data. And so we must ask for a rationale/derivation of The Standard Model within the context of the current time rather than attempt to leapfrog to a future physics without any significant experimental guidance.

---

[101] There was also a belief in that period that the speed of light was infinite.

We have two choices to explain The Standard Model. The first choice—although it seems not to be an appealing choice—is to take The Standard Model lagrangian (with perhaps a few refinements, and unification with quantum gravity) as The Fundamental Theory and simply accept its peculiarities. The other choice is to develop a reasonable set of postulates from which the known particles and interactions of nature follow as well as the peculiar form of The Standard Model lagrangian.

Recently a series of books[102] by this author have appeared that have developed a set of postulates that lead to the known particles, interactions and the form of The Standard Model lagrangian. The great virtue of this derivation, which is in fact a construction,[103] is that it leads directly to The Standard Model (possibly with a fourth generation and WIMPs – Weakly Interacting Massive Particles that may be the constituents of Dark Matter) but without a plethora of new, unseen particles.

7.3 Derivation vs. Construction of a Physical Theory

Let us assume that we have a physics theory based on a lagrangian such as that of The Standard Model that describes physical phenomena such as the accumulated experimental data on elementary particles and their interactions.

Perhaps the lagrangian in itself describes the phenomena and a deeper layer of physical theory is not possible. (A simple example of this case is a classical particle in a conservative potential.)

But in other cases it is possible that the lagrangian is the result of a more fundamental physics. (An example of this situation is the Landau-Ginzberg theory of superconductivity, which is based on the more fundamental theory of Cooper pairing.)

It was almost universally believed (until this author's work) that The Standard Model is a phenomenological theory that is a consequence of a more fundamental theory of elementary particles. The open question, of course, is the nature of the more fundamental theory. We can define a set of postulates that lead to The Standard Model. But we realize from simple logic that the ability to derive a theory from a more fundamental theory does *not* prove the more fundamental theory is the one and only correct theory.

---

[102] Blaha (2006), (2007b), (2008), (2009), (2010a) and (2010c).
[103] A construction starts with a set of axioms and proceeds to develop a theory based on the addition of features that are consistent with the set of axioms.

So, in seeking to define a more fundamental theory, that is *most likely* to be the true physical basis of an experimentally verified (phenomenological) theory, it seems reasonable to assume six principles:

1. Logic applies to metaphysical and physical derivations and discussions. (This principle is clearly assumed by all metaphysicians.)
2. In the case of several alternative choices Ockham's Razor should be used to determine the correct choice.[104]
3. We assume the universe is based on the smallest set of properties or features that lead to the greatest variety of phenomena.[105] Alternately put, we choose the most minimal conditions necessary to lead to the known phenomena of nature.
4. Whatever physical entities exist in the universe they can only make their presence known by interacting (exerting forces) with other entities. Thus each physical entity interacts with at least one other type of entity.[106]
5. The behavior and properties of physical entities must be reproducible whenever the same circumstances occur. (Reproducibility of results.) Behavior can be deterministic or can be specified in terms of probabilities. Consequently the behavior and properties must be governed directly or indirectly by mathematics.
6. We assume that space and time exist, and that the properties of entities are local in the sense that the properties of an entity depend on the point in space and time where it is located.[107]

Further, it will be seen that the derivation of a specific phenomenological theory can best be viewed as a construction from first principles rather than as a strict derivation from axioms. A *construction* begins with a set of axioms, and then develops a theory based on the axioms, and on the addition of features that are consistent with, and share the spirit of, the set of axioms.

---

[104] One very clear reason for this principle is that physics is very difficult and the simplest choice is usually the choice that enables further progress to occur. An example is Copernicus' theory that the planets circled the sun. This theory was nicely adapted to support Newton's work on the theory of gravitation. On the other hand, the Ptolemaic theory is more complex (although not inconsistent with Newton's theory of gravitation.) Copernicus' theory was far simpler than Ptolemaic theory.

[105] A version of Leibniz's Principle of Perfection.

[106] Consequently there is no knowable physical entity with no interactions with other types of entities.

[107] This principle is the basis of local rotational invariance in Yang-Mills theories or sectors of physical theories.

Euclid's geometry is an example of a construction although most students have viewed it as strictly derived from its five axioms. As logicians have pointed out the geometrical figures used to prove many geometrical theorems embody extensions of the five axioms, and are not implied by the axioms. Thus ***Euclid's geometry is a construction*** – not a strictly derived theory.

In the spirit of Euclid's approach we can construct a fundamental theory of Physics from a set of initial postulates. These postulates will be extended as we proceed by reasonable constructs that are consistent with, and reflect the spirit of the initial postulates.

7.4 What Purpose does a Construction or Derivation Serve?

As the eminent logician Frege indicated in the introductory quote there are several benefits in the reduction of a theory to a more fundamental theory (set of postulates). These benefits include:

i)   The more fundamental theory is usually simpler and more comprehensive.

ii)   The derivation of the phenomenological theory shows the origin of the various parts of the derived theory, and the interdependence of the parts of the derived theory and their derivation from a particular postulate or set of postulates.

iii) The more fundamental theory enables us to consider a deeper level of physics and perhaps find a path to a yet deeper level.

iv)   The more fundamental theory enables us to consider alternative postulate sets and the universes to which they lead. (A good example of this possibility is the controversy over Euclid's Fifth Postulate and the non-Euclidean geometries that emerged from the controversy.)

7.5 The Rigor of a Derivation

In defining a set of postulates that lead to a fundamental theory of Physics, either exactly or as an approximation, the question of the rigor of the derivation unavoidably appears. First there is the question of the rigor of differential and integral calculus which is still an issue despite the apparent rigorous development of calculus in the nineteenth century by Dedekind and others. To show the issue is still alive we simply mention the question, "If a point has no width or breadth, what does it mean to say the "next point" on a line?" This question inevitably leads to issues when one considers the definition of a derivative as a limit. Since "bare" particles are "point-like" the issue also surfaces in physics.

Secondly there is possibly the issue of a path integral formulation of a fundamental theory if it is a quantum field theory. The path integral formalism has not been put on a rigorous basis and it appears unlikely to be made rigorous for the foreseeable future.

These obstacles to a completely rigorous development mean that we must follow a procedure similar to the "traditional" physics approach of doing things "rigorously": reasonable moderation in rigor. In this approach we have the history of physics since Newton and Leibniz to support us. For approximately three hundred years physicists used the differential and integral calculus successfully. These areas of mathematics were not put on a quasi-rigorous footing in the view of many mathematicians until the mid-nineteenth century and there remains more to be done to obtain a truly rigorous calculus.

Thus a quasi-rigorous physicist's approach to proofs and derivations appears to be fully justified.

7.6 Consistency and Completeness of a Set of Postulates

Since our goal is to define a set of postulates at a deeper level from which we can derive the fundamental theory of Physics it is prudent to inquire about the consistency and completeness of the set of postulates so defined. It would be the height of hubris to believe that the set of postulates that we define completes the study of the fundamental nature of the universe(s). So we will take these postulates to be a step in the direction of deeper knowledge but realize that these postulates now become the subject of deeper theoretical and experimental investigation. There is likely to be more phenomena at higher energies that will extend the domain of particle physics beyond our current understanding. Indeed the vast majority of particle physicists expect new phenomena as higher energy accelerators appear. And the mysteries of Dark Matter, Dark Energy, and other unusual cosmological features, also suggest there is much more to learn since we have found that the very small is intimately connected to the very large.

However taking a provisional set of postulates seriously as an interim deeper theory of particle physics we have two "simple" issues to address: the consistency of the set of postulates and the completeness of the set of postulates.

The consistency issue can be resolved by two remarks:

i) The consistency of a set of postulates cannot be mathematically proved, in principle, within the framework of the theory it defines according to the celebrated Consistency Theorem of Gödel.

ii)   From a physicist's point of view a theory is consistent if it predicts a unique result for all possible experiments within the domain of applicability of the theory. This criterion is, of course, impossible to meet since the number of possible experiments is infinite. (Presumably, the predictions are verified by experiment as well.)

For example the path integral formulation of The Standard Model appears to fulfill the second consistency criteria since it provides, in principle, a unique, well-defined approach to calculating any experimental result.

The completeness of a set of postulates is determined by the theoretic results it implies. If we assume The Standard Model for example, is a complete description of particle physics (and, of course, we do not) and if the postulates lead to The Standard Model, then the postulates can be considered complete from a physical point of view. If new phenomena are found that require The Standard Model be extended, then the set of postulates will have to be extended as well. The extended set of postulates should then imply the extended Standard Model. (The possibility exists that some of the postulates might have to be modified as well.)

7.7 The Difference Between a Mathematical-Deductive System and a Fundamental Scientific Theory

The eminent philosopher and logician R. B. Braithwaite[108] has said, "The irreducible difference between the propositions of logic and mathematics and those of a natural science are that the former are logically necessary and the latter contingent."[109] This observation is true, in the author's opinion, in the case of interim scientific theories that are dependent upon experiment for clarification and further growth. However, a theory, which purports to be a complete "Theory of Everything", falls into the other category—a mathematical theory of a purely mathematical-deductive type—that needs, in principle no further experimentation and thus is strictly mathematical-deductive in nature.[110]

---

[108] Braithwaite (1960) p. 353.

[109] Braithwaite is not using "necessary" and "contingent" in the sense of modality where one distinguishes between the necessary and the contingent. Rather it appears he is using these words to distinguish between mathematics which has a logical structure that follows from fundamental logic and mathematical principles, and natural science which is provisional and deduced from experiment.

[110] It should be noted that plane geometry was a deductive theory until Euclid's five postulates were recognized as implying all the theorems of plane geometry. At this point geometry transitioned to a mathematical-deductive theory. Thus one can say Euclid's geometry is a

In Blaha (2010c) we claim to have a theory that is deeper than The Standard Model and that implies The Standard Model. This theory, framed in terms of postulates, is a step in the direction of a mathematical-deductive theory and thus is treated as such. However it is clearly not a complete "Theory of Everything" and does not pretend to be.

It is only a step—a significant step in the author's view—that obviates certain other theoretical attempts and brings us closer to a truly fundamental theory of great simplicity and depth that is based on the geometry of the universe.

"Theory of Everything" for plane geometry. The ultimate goal of elementary particle theory is to accomplish the same feat for elementary particle physics.

8. Selection Principles for Fundamental Physics Theories

8.1 The Universe of Physical Theories

Just as one can consider an infinite variety of possible physical universes, one can also consider an infinite variety of fundamental theories, one of which might hold for any particular physical universe. In the case of our universe there is a unique fundamental physical theory from which all physical phenomena ultimately derive.[111] Other universes, if such exist, may have the same fundamental theory or a different fundamental physical theory.

These considerations raise a number of questions:

1. Can different universes have different fundamental physical theories? If so, how is the fundamental physical theory of a specific universe selected?
2. If each (possible) physical universe must have the same fundamental physical theory to be physically acceptable,[112] then what are the selection criteria for this unique theory?
3. Since it is possible that two fundamental theories may hold in a universe, and yet have different forms,[113] can one show that the theories are completely equivalent?

Since we are only familiar with our universe the first question cannot be answered now – perhaps never – since the existence of other universes is unknown as is the possibility of exploring their properties.[114]

---

[111] This assertion was proven in section 2.4.

[112] Physically acceptable means at least that the postulates of the theory are consistent, and that the theory's predictions are well-defined, unique and are in agreement with experiment. The theory must predict only one result for any experiment although the predicted result may be stated in the form of a probability distribution.

[113] This situation occurred in the early days of Quantum Theory when both the Schrödinger wave mechanics formulation and the Heisenberg formulation of quantum theory both described quantum mechanics. These formulations were soon shown to be equivalent.

[114] The word universe is a bit vague when it is used in the physics and metaphysics literature. One reads of worm holes between universes, quantum tunneling between universes, and so on. These concepts and gedanken (thought) experiment discussions raise the question – if one can connect, or communicate with, other "universes" are they really separate universes or are they parts of the same universe with difficult paths between them? We will therefore define a *universe* to be specifically a completely self-contained entity totally unconnected with any other universe entity in any way whatsoever. Thus beings within a universe cannot communicate/contact/obtain

The second question might be answerable for our universe. One possible answer, in which the selection criteria is based on a particular formulation of Logic called Operator Logic, is described in Blaha (2010b) and (2010c). There may be other possible answers but our current understanding of Physics is not capable of providing it.

The third question is easily answered. Since each theory must make the same physical predictions for any experiment, there must be a map between the fundamental theories that establishes their equivalence.

8.1.1 Types of Fundamental Theories

There appear to be two types of fundamental physical theories:

1. Fundamental theories based on a lagrangian using the apparatus of quantum field theory and path integrals.

2. Fundamental theories based on a set of postulates from which a lagrangian formalism may, or may not, result.

An example of a fundamental theory of the first type would be a unified lagrangian theory consisting of The Standard Model and Quantum Gravity. An example of the second type of theory is presented in Blaha (2010c) in which the lagrangian of The Standard Model is derived/constructed from a set of fundamental postulates. A renormalizable Quantum Gravity is constructed from General Relativity and directly united with The Standard Model to form a complete unified fundamental theory. Another fundamental theory of the second type may be M-theory, a sophisticated generalization of string theory. M-theory has been under serious study for fifteen years but it is incomplete, lacks predictive abilities, and does not have any currently testable experimental predictions.

8.2 Possible Selection Principles for "The" Physical Theory of a Universe

There are a number of possible selection principles that could determine the fundamental physical theory of the universe:

---

knowledge of, or even be aware of, other possible universes either physically or conceptually. If communication/data acquisition/physical effects of "another" universe take place then our universe and the other universe are really distant parts of the same large universe. Beings within a universe can speculate on possible other universes and their features. These other universes are "universes of the mind" – assemblages of thought and logic created in imagination.

1. A set of fundamental space-time and internal symmetries in a specified space-time with a specified set of dimensions. A unified quantum field theory would then be constructed based on these symmetries. Possible Example: M-Theory.

2. A selection principle based on a general abstract principle such as Leibniz's minimax principle – *the laws of nature are of maximal simplicity yet capable of producing a universe of maximal complexity.*[115] Not feasible given the current state of physical knowledge. The combined concepts of maximal simplicity and maximal physical complexity cannot be expressed mathematically.

3. A selection principle based on a set of fundamental postulates. Example: Blaha's derivation of The Standard Model. Blaha (2010c)

4. A path integral formulation of the universe of possible lagrangians with a variational principle that selects the fundamental lagrangian of the universe. Case 2 is a special case of this case.

5. The fundamental theory of the universe might be selected by a set of boundary conditions/initial conditions through a mechanism unknown to us. These boundary conditions would somehow set the physical laws and constants of the universe. Since a knowledge of this type of mechanism does not exist, nothing more can be said at the present time. However, it does exist as a logical possibility.

6. The fundamental theory of physics might follow directly from fundamental Logic principles supplemented by physical postulates such as the validity of quantum field theory and the path integral formulation. This is a special case of case 3.

Of all of the above possibilities it appears that case 6 is the only case that can be successfully considered. As Blaha (2010c) shows it can lead to a successful derivation of form of The Standard Model. This is the only approach that can make that claim.

8.3 A Yet Deeper Level?

The above considerations raise the hope that an ultimate end will be found to our search for knowledge of our universe – the hope that we

---

[115] In Leibniz's words, "at the same time the simplest in hypotheses [i.e., its laws] and the richest in phenomena." Quoted in Rescher (1967) p.19.

will know all there is to know of Physical Reality. Yet upon consideration it seems that such a final end to our striving for knowledge of Reality will not take place.

The lack of finality follows when we consider that all "fully defined" mathematical-deductive systems must in the end rely on undefined terms – primitive terms – which we can only partially grasp and only partially identify with the entities and properties of Reality. Thus we wind up mathematizing the primitive terms, and their relations, but their essential Reality eludes us.

We can only conclude that the relationships between entities expressed in our theories is the only Reality that we can truly understand. "That is all we [can] know and all we need to know!"

9. The Only Logical Choice for a Physical Selection Principle - Logic

9.1 Essentiality of Logic

Classical Metaphysics is largely composed of attempts to define fundamental principles, to clarify concepts and to categorize entities. The key basis of these attempts is Logic. This was pointed out many years ago by such philosophers as Leibniz.[116] Logic enables philosophers to use fundamental principles to derive new principles. Logic enables philosophers to clarify concepts and their differences. Logic is used to categorize and differentiate between entities.

Thus Logic is the engine of Philosophy and Metaphysics as Reality is the meat of Metaphysics.

In considering Metaphysics Leibniz felt that propositions of the subject-predicate form or reducible to a combination of clauses of the subject-predicate form, sufficed to meet the needs of metaphysical discussion and analysis. This view has been disputed by some metaphysicians but from the viewpoint of physical Reality it appears the view of Leibniz is correct.

Consequently we will consider (physical?) Logic as applying to axioms and propositions that have subject-predicate form or are reducible to subject-predicate form.

We will relate Logical propositions to experiments on physical Reality. Subsequently we will use Operator Logic to outline a derivation/construction of physical Reality in a manner reminiscent of Plato's Theory of Ideas and Reality, and their connection through mathematics.

9.2 Logic and Quantum Theory

9.2.1 A Logic View of Physical Experiments

At first glance experiments on physical Reality have no direct connection to Logic. However as Blaha (2009), and his earlier work, showed a quantum experiment (specifying intermediate stages of the experiment and a specific result) can be stated as a proposition in subject-predicate format, or as a proposition consisting of a set of clauses in subject-predicate format. Thus quantum experiments can be mapped

---

[116] Rescher (1967) pp. 22-23.

in a one-to-one fashion with propositions. We describe the form of the map in the next section.

9.2.2 Filtration Stages in a Quantum Experiment and Their Mapping to the Form of a Proposition

If we consider an example of a proposition such as

The car is a small, green Ford.

we see that the proposition is equivalent to

The car is small, and the car is green, and the car is a Ford.

Now if we consider the set of all cars then we see that the first clause restricts the set of all cars to the subset of small cars. The second clause restricts the subset of small cars to green cars, and the clause further restricts the subset to the yet smaller subset of Ford cars. Each restriction to a yet smaller subset we can call a *filtration* in the sense that each restricting clause filters a set to a smaller subset.

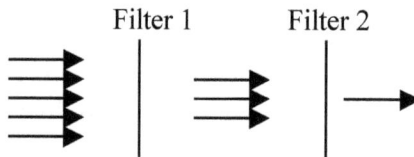

Figure 9.1. A two filter experiment that selects particles with the exact velocity of 100 miles per second from a stream of incoming particles with a variety of velocities.

Now let us consider a (marginally) quantum experiment[117] that progressively filters a stream of particles that have a range of velocities so that only particles with a velocity of 100 miles per second emerge. Let us assume two filters: filter 1 only lets particles with velocities greater than or equal to 100 miles per second pass through, filter 2 only lets particles less than or equal to 100 miles per second pass through. As a result only particles with a velocity exactly equal to 100 miles per second pass through the two filters. Fig. 9.1 illustrates this simple experiment.

This experiment can be expressed as a proposition:[118]

---

[117] The filters are assumed quantum in nature.
[118] We use quotes "…" to indicate subjects and predicates.

A "stream of particles" "passes through a filter that allows only particles with velocity greater than or equal to 100 miles per second through" and then "passes through a filter that only allows particles of velocity less than or equal to 100 miles per second through" so that only "particles with a velocity that is exactly 100 miles per second" "emerge."

Or, in short,

"A stream of particles with a range of velocities" "is reduced to a stream of particles with a velocity of 100 miles per second."

9.2.3 Conceptual Correspondence Between Logic and Quantum Theory

Clearly there is a conceptual correspondence between logical propositions and experimental propositions. In both cases the predicates can be described as filters that perform a selection upon the set specified by the subject, or by the implied set of which the subject is generally a member. In the above logic proposition, the subject begins as a member of the set of all cars and is then filtered to a very specific subset of the set of all cars.

Therefore it is clear that we can view subject-predicate logic propositions and their generalizations to subject-predicate clauses as a series of filtrations similar to the filtrations that take place in quantum physics experiments. We define this feature as the fifth principle of Relativistic Quantum Metaphysics:

5. Subject-Predicate propositions in Logic can be viewed as filtrations analogous to filtrations that take place in quantum physics experiments.

With this concept as a beginning point we developed a new form of Logic that we called Operator Logic[119] that resolved known logical paradoxes, including Gödel's Undecidability Theorem. It also furnishes a starting point for a transition from the world of Ideas (Logic propositions) to the world of Reality. Pursuing this line of development we eventually derived the form of The Standard Model of Elementary Particles.[120]

Thus we view Logic as the origin of the fundamental theory of Physics (case 6 in chapter 8). In a sense this view could not be more satisfactory. Logic in its primary form with two logical values for any proposition – true or false – is both unique and also critical for the

---

[119] Blaha (2010b).
[120] Blaha (2010c) and earlier books.

development of any mathematical-deductive theory. Without Logic physics and mathematics are not possible. In the following sections we will outline the derivation/construction of The Standard Model of Elementary Particles in non-mathematical terms. The complete mathematical treatment is presented in Blaha (2010c).

9.3 Formulation of Logic & Particle Spin

In Logic a proposition can be either true or false.[121] In physics the spin ½ elementary particles, the leptons and quarks (Table 4.1), can be in "spin up" or "spin down" states. The double-valuedness of logical propositions, and of spin ½ particles, enables a mathematical map to be created between a mathematical formulation of Logic and spin ½ particles.

This map is global in the sense that the location of a logical proposition and a spin ½ particle are not initially specified. However realizing that the truth of logical propositions is local in general: what is true in one place may well be false in another; and that the state of a particle may vary from place to place; we introduce space-time coordinates that specify the local of a proposition or particle.

The introduction of space-time coordinates for particles requires a specification of the transformation laws for observers in different coordinate systems.[122] Imagine two observers moving at a relative speed v. When they each measure the space and time locations of a point they will obtain different space and time coordinates related by the Theory of Relativity – this is especially evident when the relative speed v gets very large and approaches the speed of light.

So in attaching a space-time location to a particle we require it be specified in a way that is consistent with the Special Theory of Relativity.

Consequently we take the form of a spin ½ particle state, and attach relativistic coordinates. The result is mathematical expression that looks like the wave function of a Dirac particle at rest. We then transform that expression to that of a moving spin ½ Dirac particle.

After further mathematical manipulations we find:

1. There are four general types of spin ½ particles.
2. Two types correspond to the two types of leptons.
3. Two types correspond to the two types of quarks.

---

[121] Or undecidable. This case is not relevant to the current discussion.

[122] This is also true in general for specifications of location in logic propositions but in practice only simple specifications of location tend to be considered. So the impact of general spce-time considerations is ignored.

4. Each type of quark has three subtypes.
5. The eight kinds of particles described by items 1 through 4 are duplicated four times to generate the fermion generations. The three known generations are listed in Table 4.1 of The Standard Model.

Thus starting from Operator Logic we can construct the known spectrum of fermions of Table 4.1 giving us a mathematical path from Operator Logic to the basic equations of matter.

The Platonists, who first conceived the concept of a mathematical path from the Realm of Ideas to the Realm of Reality, of course had no knowledge of Operator Logic, or The Standard Model, or the mathematical path between them. But they were able to develop a conceptual vision of the Realm of Ideas and a conceptual mathematical bridge to Reality.

9.4 Space-Time: Sub-Light and Superluminal

An essential part of the derivation/construction from Logic to The Standard Model was the extension of the transformations in the Theory of Special Relativity to include superluminal transformations: transformations between coordinate systems moving at relative velocities larger than the speed of light. Such transformations are not excluded by physics except for transformations where the relative velocity is exactly equal to the speed of light. The form of the spectrum of fermions (Table 4.1) and the group structure of the ElectroWeak interactions $SU(2) \otimes U(1) \otimes U(1)$ and the Strong Interactions $SU(3)$ emerge directly from the mathematics of superluminal transformations. This author believes that we will observe superluminal transformation phenomena in nature in the future. Thus the derivation of The Standard Model has opened a new doorway into Reality.

9.5 The Necessity of Time, and an Arrow of Time, in Logic

A number of logicians have noted that a concept of time is implicit in conventional Logic. For example, proofs of theorems proceed step by step from initial postulates and theorems to a theorem's proof. Embedded in that process is a notion of discrete time steps, and a direction of the time steps. The directionality of the time steps[123] specifies an "arrow of time." The question of the Arrow of Time – why time proceeds forward and not backward – has been a subject of much

---

[123] After all one does not proceed "backward" from a theorem through the proof steps to the initial postulats.

discussion over the years. In the present situation Logic and Operator Logic automatically embody an arrow of time.

Not only is this true for proofs but it is also true for statements. Although the order of the parts of a statement is language dependent the order is usually specific and consecutive within a given language and thus has a time order as well.

So we conclude that Logic embodies discrete time steps and a definite concept of time ordering – an Arrow of Time.

Having ascertained that discrete time, and time ordering, is implicit in Logic we now define physical time as the continuous limit of discrete time (with the understanding that physical time may be discrete and may possibly consist of very small time steps of time intervals of the order of the Planck time scale 5.39×10^{-44} seconds.) Discrete time intervals of that order of magnitude are not detectable experimentally at present or in the foreseeable future. Thus the assumption of continuous time, with an arrow of time, is satisfactory.

9.6 Why add Space to Logic?

Space is necessarily a part of the Realm of Ideas because propositions often depend on a spatial location. Thus we must add spatial (and time) dimensions to our specification of the Realm of Reality as well as the Realm of Ideas.

The number of space-time dimensions is an important issue. Clearly if spin ½ particles exist in Reality, as we know they do, then they must be "spinning" in spatial dimensions. The number of components of a spin ½ particle is related to the total number of time and space dimensions.[124] Based on physics formulas we find the results in the following table.

Total Number of Space-Time Dimensions d	Number of Spin 1/2 Components
1	1
2	2
3	2
4	4

Table 9.2. The number of spin ½ components for various numbers of space-time dimensions.

---

[124] Weinberg (1995) p. 216 points out: For the case of an even number of dimensions d a spinor has $2^{d/2}$ components. For the case of an odd number of dimensions d a spinor has $2^{(d-1)/2}$ components.

The case of d = 1 is immediately ruled out because spin ½ particles have a minimum of two components. The case d = 2 is also ruled out because spin ½ particles in a one-dimensional space reduce to scalar particles (particles without spin), and Reality has true spin ½ particles. The case d = 3 is ruled out because in two spatial dimensions there is no difference between left-handedness and right-handedness. But handedness is an important feature of physical Reality. Thus the minimal number of spatial dimensions that yield true physical spin ½ particles and support "handedness" is three spatial dimensions. This case meets Leibniz's criteria: principle 2 of chapter 8. The "simplest" features associated with space are spin and handedness. They yield a rich spectrum of particle types and interaction types (maximal complexity).

Thus we have a rationale for the extension of Operator Logic (and physical Reality) to include one time and three spatial dimensions.

Appendix 9-A Approaches to a Deeper Level of Reality

While we have shown that Logic furnishes a basis for the development of The Standard Model of Particles, the additional assumptions – what we call the Knowledge Base (discussed below)– remain to be determined. It may be that, like Euclidean geometry, we can go no further than a set of postulates/axioms (assumptions). The number of explicit axioms needed to obtain The Standard Model in Blaha (2008) was twenty-three.[125] There were also a number of implicit assumptions in the derivation in Blaha (2008) just as there are a number of implicit assumptions in Euclid's'geometry which are only evident when figures are drawn in the process of proving theorems.

Irrespective of the question of the number and content of the axioms required to prove The Standard Model of Elementary Particles and thus establish the basis of Reality as we know it, the question that immediately arises is the source/reason for these particular axioms.

The possible sources for the axioms leading to The Standard Model are:

1. There is no reason. They are just the basis of Reality.
2. They follow from an unknown deeper unifying principle(s).
3. They follow from a deeper known theory such as Superstring Theory.
4. They follow from an unknown mechanism that establishes order in the form of the axioms from chaos. A mechanism or explanation for the persistence of such order over billions of years must also be found.
5. They follow because they are the only totally consistent set of physical axioms.
6. They are the result of chance and one of many possible sets of axioms.
7. They follow because they are required for life, as we know it, to exist. (The Anthropic Principle) This case is clearly a subcase of cases 1 and 6.

---

[125] The number of postulates in the Complexon Standard Model in Blaha (2010c) is 29.

While case 1 is possible, it is intellectually unsatisfying and so we will not pursue it. We will discuss case 2 in the following subsections. We will not discuss case 3 and refer the reader to the physics literature.

Case 4 seems unlikely to the author because of the secondary requirement that order persists indefinitely.

Case 5 brings us to Gödel's Consistency Theorem, which roughly states that a set of axioms cannot be proved to be consistent within the mathematical-deductive system that they define. Thus case 5 is unprovable.

Case 6 introduces an infinity of possible choices for the set of axioms for physics. The broadness of the set of choices makes this an unattractive possibility.

Case 7 is certainly true for humanity but recent studies have shown there is an extremely wide range of environments that can support life, and possibly, intelligent life. Thus, unless one posits religious reasons, it is not particularly clear that there is a compelling case for the Anthropic Principle. Case 7 then degenerates to case 6 – mere chance. And mere chance is not susceptible to intelligent discussion unless one can show that there is a "probability distribution" for sets of axioms and the set of axioms that governs our universe is amongst the most likely sets of axioms. Then case 7 becomes a subcase of case 2.

Therefore we find that case 2 is the most attractive case to study although case 3 (discussed elsewhere) is also of interest.

9-A.1 The Knowledge Base of Reality

As we have seen, a universe of discourse consists of a set of axioms written in terms of undefined primitive terms and the theorems derived therefrom. A *semantic* universe of discourse attributes meanings/interpretations to the primitive terms thereby giving meaning to the axioms and consequent theorems. We will call the set of primitive term definitions and the set of axioms the *knowledge base* of the semantic universe of discourse of Reality.

A knowledge base for physical Reality (at least the part of it with which we are familiar) was first expounded in a series of books by Blaha culminating in Blaha (2010c). Blaha (2008) lists a set of axioms that lead directly, and *exactly*, to most of the established features of the form of The Standard Model of Elementary Particles. Blaha (2010c) derives all of the known features of the form of the Standard Model – particularly the SU(3) color Strong interactions and SU(2)⊗U(1)⊗U(1) ElectroWeak interactions. The major known features of this theory of elementary particles were determined in a period of forty-five years (1930 – 1975 approximately). These features include parity violation, peculiar symmetries, and the somewhat complicated nature of the particle spectrum. These features are *exact* results of the axioms in Blaha (2010c), and earlier books, unlike other theoretic attempts to explain The Standard Model. Other attempts view The Standard Model as a low energy approximation of a larger theory and have no inherent justification for parity violation but rather define their theories to incorporate it.

In Blaha's derivation every basic particle of matter, whether quark or lepton, has spin ½. In Blaha (2010c) the basic algebra of Operator Logic eigenvalue operators, and that of the raising and lowering operators, is the same as the algebra of free spin ½ particles. Operator Logic is part of the Realm of Ideas. Our goal was to build a theory of Reality on the scaffolding of Operator Logic. We believe we succeeded. In building this theory we have implemented the Platonic goal of mathematically connecting the Realm of Ideas and the Realm of Reality.

Appendix 9-B. Role of the Observer in the Realm of Reality

An important issue is whether the Realm of Ideas exists independently of our knowledge of it and its contents. One might also ask the same question of the Realm of Reality: Do events happening in Reality, and the manner in which they occur, require an Observer to detect them? Or do events in Reality happen irrespective of our presence? The answer to these questions lies in Quantum Theory.

In Quantum Theory there is an implicit Observer who "sets up" experiments, performs eigenvalue measurements (filters) at various stages in an experiment, and measures the end result(s) of an experiment.

9-B.1 The Observer in Operator Logic & Quantum Operator Logic

In Operator Logic and Quantum Operator Logic an implicit "observer" (a logician or theorist) determines the truth (semantic universe of discourse) or provability (calculus universe of discourse) of a statement.

An observer has three major roles in Operator Logic and Quantum Operator Logic:

1. Create statements.
2. Determine the truth (provability) of statements.
3. Create derivations of statements (theorems).

The effect of the observer on the truth (provability) of statements depends on whether the statement is a c-number statement (Operator Logic) or a q-number statement (Quantum Operator Logic).

In the case of Operator Logic the observer has no effect on the truth value (true or false, provable or not, or probability of truth) of the statement.

In the case of Quantum Operator Logic, if the observer determines the state of a proposition (statement) at an intermediate point(s) in the proposition (statement) then the probability of the statement is affected.

Thus the Operator Logics, like Quantum theory, have an observer although the observer is usually implicit. We therefore conclude that the

Realm of Ideas does not exist independently of our knowledge of it, but depends on an Observer – the Philosopher or Logician.

10. Being and Existence of the Material World

Having established the Ideas of Quantum Theory, Time and Space and their counterparts in Reality we now turn to the Idea of Being.

The question of being or existence has been a subject of discussion in Philosophy and Metaphysics for millennia. In the absence of "experimental" information the discussions have centered on the definitions of being and the implications of these definitions for the "properties" of being.

For many scholars the state of Philosophy and Metaphysics was considered satisfactory in the 20ᵗʰ century with respect to issues such as Being. For example, Hans Bethe, perhaps the dominant US figure in theoretical physics from the 1930's through the 1950's, and a Nobel Prize winner, stated that at the beginning of his "graduate" studies in the 1920's he considered the state of Philosophy and Metaphysics, and concluded that they were satisfactory. He then decided to become a physicist where he felt that he could make significant contributions (which he did by discovering the solar energy carbon cycle, and making notable contributions to nuclear physics and quantum field theory as well as guiding a generation of physicists including R. P. Feynman and M. Gell-Mann). In our view, Professor Bethe's views on Metaphysics did not anticipate the impact of Quantum Theory and Relativity.

In this chapter we will consider a phenomenological (Realistic) view of Being and nothingness based on the experimental observations of the creation and annihilation of particles over the past eighty years. The experimental picture that we will use dates from the 1930's when the occurrence of particle creation and annihilation was first recognized. The mathematics of quantum field theory adequately describes particle creation and annihilation as seen experimentally. Since particle creation is the "creation" of Being and since particle annihilation is the "destruction" of Being one could say that the issues of Being and non-Being can be resolved by experimental observations and their theoretical analysis – thus making metaphysical speculations experimentally accountable. *Experiment can provide direct guidance on the nature of Being.*

Having said that, we recognize that the mathematics of creation and annihilation of particles somehow doesn't fully answer the question, "What is Being?" from a human perspective. Part of the problem is that we don't exactly "know" what a particle is. We know particles have particle-like properties and wave-like properties as well. But statements about properties do not address the issue of what Being is in itself. Can we say particles are composed of a substance or substances? If so, what substance or substances? Or are particles merely form without substance?

Earlier in sections 6.2 and 6.3 we described the view of substances that emerges from Physics: the universe, and all particles within it, is composed of one substance. Particles consist of this substance as does the vacuum. Each particle has a distinctive form that we characterize by its properties: its space-time properties and its internal quantum numbers. Thus we have a mathematical knowledge of being.

10.1 Origin of Being

The components of Reality are tangible—perceivable by our senses (directly or indirectly), and the components of Reality interact with each other. But how did the components of Reality acquire existence?

Remarkably, there is strong evidence that the universe emerged from a point (or a "small" neighborhood of a point) at its beginning, The Big Bang. So the question of Being may degenerate to the question of "Being" at a single point without extension, without space or time, and perhaps without content. A number of theorists have suggested that the Big Bang is a quantum fluctuation—something emerging from nothing (the "vacuum") in such a way that the sum total of the emergent fluctuation has zero energy and thus is still nothing if considered *in toto*.[126]

In view of the major uncertainties, both physically and philosophically, in the understanding of existence and Being the question of the origin of Being is primarily work for the future. This chapter provided a preliminary discussion of Being based on our knowledge of particle physics – a subject which confronts existence and non-existence (Being and non-Being) on an everyday experimental and theoretical basis.

---

[126] Being then becomes an illusory artifact of our consciousness. Combinations of forms (particles) constitute matter and the forces between these particles en masse give matter its solidity or liquidity as the case may be. The forces between chunks of matter when they collide are the cumulative effect of the forces between the particles constituting each chunk.

11. The Platonic Conception of the Relation of Ideas to Reality

Plato, and subsequent Platonic philosophers, postulated that there existed a Realm of Ideas independent of the human mind, consisting of the concepts and thought processes (Logic) existent in the mind.[127] Each thing, both material and conceptual, in Reality had an abstract counterpart in the Realm of Ideas that embodied its features. Thus there was an Idea of a plow, an Idea of Justice, and so on. The Realm of Ideas was connected through mathematics (Number) to the Realm of Reality with which we are familiar. The nature of the connection was not known to the Platonic philosophers since the relevant mathematics and physics were not known.

Plato, and subsequent philosophers, were of course not familiar with the modern view of Reality, which has only become apparent in the quantum revolution of the twentieth century. The concepts of Quantum physics apply to non-relativistic quantum mechanics and relativistic Quantum Field Theory.

Quantum Theory has a formal method of viewing the world. It describes the process of quantum experimental measurement as a series of filtrations (or refinements that select certain properties or states) that occur at various stages of an experiment. Since all observations are experiments, it applies in principle to all the ways that we obtain information about the universe in which we live. In the macroscopic world quantum effects are usually negligible. In the very small, quantum effects often dominate physical processes. Quantum Theory applies in principle in all cases.

So the gropings of Plato, and other philosophers, towards a total view of Reality had an element of truth in that there is a fundamental set of Ideas that govern Reality, but was also wrong—for good reason—Reality does not map directly to Ideas in the one-to-one fashion envisioned by the Platonists.

Instead, the set of Ideas is the set of laws of physics: Quantum Theory, the Theory of Relativity and elementary particle theory. The phenomena of Reality behave according to the laws of physics. The mathematics connecting Ideas to Reality has been shown to start from

---

[127] Plato discusses the concept of Ideas in the dialogue Parmedides as well as in other dialogues.

Logic and end in the form of The Standard Model of Elementary Particles. Thus Platonic thought was correct in a general sense; but the specifics were incorrect due to the primitive state of physical science until recently.

11.1 Operator Logic and Quantum Operator Logic Exist Independent of Our Knowledge of Them

We have constructed Operator Logic and Quantum Operator Logic[128] in the "language" of linear vector spaces, and, more specifically, based on the view that a proposition is analogous to a quantum experiment in that it proceeds to particularize subjects in the same manner that a quantum experiment filters quantum states to achieve a particular final state. Linear vector spaces appear in many areas of physical Reality. Since Reality cannot be internally inconsistent, incomplete or "incorrect", Operator Logics are well founded and exist independently of our knowledge of them.[129] Our understanding of quantum physics only emerged in the twentieth century yet the universe has followed the laws of quantum theory since its beginning. Thus we must attribute a conceptual reality to the concepts and laws of Quantum Theory. The extreme precision of the theory of Quantum Electrodynamics strongly indicates that Quantum Theory is correct.

Consequently the only reasonable philosophical stance is Platonist—the concepts and laws of Quantum Theory have a true existence outside of our knowledge of them.[130]

11.2 The Realm of Ideas

The preceding section shows that our formalisms for Operator Logic and Quantum Operator Logic exist independently of our awareness of them. One may ask what other Ideas exist in the Realm of Ideas. Confining ourselves to the physical, we see that the Realm of Ideas must contain Operator Logic, Quantum Operator Logic, Mathematics, and the fundamental theory of physical Reality. The consideration of other possible parts of the Realm of Ideas – Ethics and Theology – are beyond the scope of this book.

---

[128] Blaha (2010b).

[129] The deepest part of our current physical understanding of the universe is Quantum Theory. All branches of physics ultimately conform to it. It is extremely unlikely that Quantum Theory will change.

[130] This statement should not be confused with the quantum mechanical requirement of an observer to make quantum mechanical measurements. Quantum physical processes can proceed without an observer although experimental results, being probabilistic, can only be determined by an observer.

12. Logic, Language and the Universe

We have seen that physical Reality can be viewed as based ultimately on Logic – the only viable underlying framework that leads to The Standard Model (sections 9.3 and 9.4).

Logic, itself, is ultimately based on language – human languages and mathematical languages. Otherwise propositions could not be framed.

Interestingly the particles of The Standard Model numbering currently 36 can be viewed as letters in an alphabet that transform into one another through particle interactions. Quantities of matter then can be viewed as words. Going one step further one can view the universe as one enormous word extending from the beginning of time to the end of time. Thus we come to the view that language is the essence of Reality and the one substance composing all things is the embodiment of language – mathematically complicated and yet conceptually understandable. These views have been considered at length in our earlier work.[131]

12.1 Logical Equivalence of Languages

All logic is expressed in human or symbolic languages. As a result there is an intimate connection between logic and language. Statements and deductive systems require a sufficiently robust language to express their content. An adequate language must have the equivalent of predicates, subjects, connectives and quantifiers.

A question of some interest is the equivalence of languages. When are two languages equally capable of expressing the statements of a universe of discourse? Clearly they must have sets of equivalent terms although a term in one language might be a combination of terms in the other. However in some cases of human languages this type of equivalence is hard to achieve. A classic example is the Greek language in comparison to English. The Greek language has hundreds of words expressing various forms and nuances of love while English, in comparison, has few words for love. In other areas of the world some languages have a plethora of words for one aspect of nature, or another, which have no simple analogue in European languages.

---

[131] Blaha (2002), (2005b) and (2005c).

One might think that the equivalence of languages is not of great importance. However, the growth of culture and science is directly tied to the growth in their terminology and the concepts that they embody. An example is the growth in the knowledge of quantum theory in the twentieth century, which introduced a host of new terms in physics. Thus the equivalence of languages reflects, to some degree, the equivalence of the range of universes of discourse (and their intellectual content) that the languages can support.

The requirements for two languages to be equivalent are:

1. Equivalent expressions in the two languages must have the same truth value.

2. Any expression in one language must have an expression with equivalent meaning in the other language.

3. The primitive terms of one language must be equivalent to the primitive terms in the other language, or to combinations of the primitive terms in the other language.

12.2 Operator Languages

Hitherto the languages that have been considered in studies of Logic have been human languages or symbolic languages that we will call c-number languages following the terminology of Quantum Theory. A c-number is a quantity, a number, character, or string of characters, that is not an operator. We have introduced the concept of a q-number language for logic in the definition of Quantum Operator Logic. Primitive terms are represented by operators in a linear vector space. This q-number language is as valid as c-number languages to express statements.

Q-number languages have the added advantages of

1. Being able to project out undecidable statements and thus resolving the profound issues raised by the Gödel Undecidability Theorem. C-number formulations of logic are an inadequate framework. Q-number logic is the proper framework for Logic.

2. Enabling the creation of subuniverses of discourses, direct products of universes of discourse, and direct sums of universes of discourse.

3. Furnishing a unifying framework for deterministic logic and quantum probabilistic logic that is guaranteed to be well-formed and consistent since it is based on Quantum Measurement Theory, which, embodying Reality, cannot be inconsistent or incomplete.

4. Providing matrix formulations of Operator Logic and Quantum Operator Logic.

5. Implementing the Platonic concept of Ideas connected mathematically to Reality.

12.3 Quantum Languages, Grammar, Turing Machines, Computers, and Computer Programs

Blaha (2010b) develops Operator Logic and Quantum Operator Logic. Blaha (2005b) developed the concepts of Quantum Languages, Quantum Grammars, Quantum Turing Machines, Quantum Computers, and Quantum Computer Programs; and proved Gödel's Undecidability Theorem required the fundamental laws of Nature to be quantum.

These books complement each other by bringing the Quantum concept, which is undoubtedly the deepest knowledge that we have of Reality, to Logic and Language.

A reading of Blaha (2005b) displays a remarkable similarity in the concepts and mathematics of quantum languages and our development of Quantum Operator Logic.

Together they give us a coherent weltanschauung of Thought and Reality. The major open question is the determination of the knowledge base of Reality. In this author's view this question will be resolved by an extension of our understanding of the formulation of the nature of space, time, and substance within the framework of Quantum Operator Logic and Quantum Language.

REFERENCES

Akhiezer, N. I., Frink, A. H. (tr), 1962, *The Calculus of Variations* (Blaisdell Publishing, New York, 1962).

Bjorken, J. D., Drell, S. D., 1964, *Relativistic Quantum Mechanics* (McGraw-Hill, New York, 1965).

Bjorken, J. D., Drell, S. D., 1965, *Relativistic Quantum Fields* (McGraw-Hill, New York, 1965).

Blaha, S., 1995, *C++ for Professional Programming* (International Thomson Publishing, Boston, 1995).

_____, 1998, *Cosmos and Consciousness* (Pingree-Hill Publishing, Auburn, NH, 1998 and 2002).

_____, 2002, *A Finite Unified Quantum Field Theory of the Elementary Particle Standard Model and Quantum Gravity Based on New Quantum Dimensions™ & a New Paradigm in the Calculus of Variations* (Pingree-Hill Publishing, Auburn, NH, 2002).

_____, 2004, *Quantum Big Bang Cosmology: Complex Space-time General Relativity, Quantum Coordinates,™ Dodecahedral Universe, Inflation, and New Spin 0, ½, 1 & 2 Tachyons & Imagyons* (Pingree-Hill Publishing, Auburn, NH, 2004).

_____, 2005a, *Quantum Theory of the Third Kind: A New Type of Divergence-free Quantum Field Theory Supporting a Unified Standard Model of Elementary Particles and Quantum Gravity based on a New Method in the Calculus of Variations* (Pingree-Hill Publishing, Auburn, NH, 2005).

_____, 2005b, *The Metatheory of Physics Theories, and the Theory of Everything as a Quantum Computer Language* (Pingree-Hill Publishing, Auburn, NH, 2005).

_____, 2005c, *The Equivalence of Elementary Particle Theories and Computer Languages: Quantum Computers, Turing Machines, Standard Model, Superstring Theory, and a Proof that Gödel's Theorem Implies Nature Must Be Quantum* (Pingree-Hill Publishing, Auburn, NH, 2005).

_____, 2006a, *The Foundation of the Forces of Nature* (Pingree-Hill Publishing, Auburn, NH, 2006).

_____, 2006b, *A Derivation of ElectroWeak Theory based on an Extension of Special Relativity; Black Hole Tachyons; & Tachyons of Any Spin.* (Pingree-Hill Publishing, Auburn, NH, 2006).

_____, 2007a, *Physics Beyond the Light Barrier: The Source of Parity Violation, Tachyons, and A Derivation of Standard Model Features* (Pingree-Hill Publishing, Auburn, NH, 2007).

_____, 2007b, *The Origin of the Standard Model: The Genesis of Four Quark and Lepton Species, Parity Violation, the ElectroWeak Sector, Color SU(3), Three Visible Generations of Fermions, and One Generation of Dark Matter with Dark Energy* (Pingree-Hill Publishing, Auburn, NH, 2007).

_____, 2008a, *A Direct Derivation of the Form of the Standard Model From GL(16) (Pingree-Hill Publishing, Auburn, NH, 2008).*

_____, 2008b, *A Complete Derivation of the Form of the Standard Model With a New Method to Generate Particle Masses Second Edition* (Pingree-Hill Publishing, Auburn, NH, 2008)

_____, 2009, *The Algebra of Thought & Reality: The Mathematical Basis for Plato's Theory of Ideas, and Reality Extended to Include A Priori Observers and Space-Time Second Edition* (Pingree-Hill Publishing, Auburn, NH, 2009).

_____, 2010a, *Operator Metaphysics: A New Metaphysics Based on a New Operator Logic and a New Quantum Operator Logic that Lead to a Mathematical Basis for Plato's Theory of Ideas and Reality* (Pingree-Hill Publishing, Auburn, NH, 2010).

_____, 2010b, *The Standard Model's Form Derived from Operator Logic, Superluminal Transformations and GL(16)* (Pingree-Hill Publishing, Auburn, NH, 2010).

_____, 2010c, *SuperCivilizations: Civilizations as Superorganisms* (McMann-Fisher Publishing, Auburn, NH, 2010).

_____, 2011a, *21st Century Natural Philosophy Of Ultimate Physical Reality* (McMann-Fisher Publishing, Auburn, NH, 2011).

_____, 2011b, *All the Universe! Faster Than Light Tachyon Quark Starships & Particle Accelerators with the LHC as a Prototype Starship Drive Scientific Edition* (Pingree-Hill Publishing, Auburn, NH, 2011).

_____, 2011c, *From Asynchronous Logic to The Standard Model to Superflight to the Stars* (Blaha Research, Auburn, NH, 2011).

_____, 2012a, *From Asynchronous Logic to The Standard Model to Superflight to the Stars volume 2: Superluminal CP and CPT, U(4) Complex General Relativity and The Standard Model, Complex Vierbein General Relativity, Kinetic Theory, Thermodynamics* (Blaha Research, Auburn, NH, 2012).

_____, 2012b, *Standard Model Symmetries, And Four And Sixteen Dimension Complex Relativity; The Origin Of Higgs Mass Terms* (Blaha Reasearch, Auburn, NH, 2012).

_____, 2013a, *Multi-Stage Space Guns, Micro-Pulse Nuclear Rockets, and Faster-Than-Light Quark-Gluon Ion Drive Starships* (Blaha Research, Auburn, NH, 2013).

_____, 2013b, *The Bridge to Dark Matter; A New Sister Universe; Dark Energy; Inflatons; Quantum Big Bang; Superluminal Physics; An Extended Standard Model Based on Geometry* (Blaha Reasearch, Auburn, NH, 2013).

_____, 2014a, *Universes and Megaverses: From a New Standard Model to a Physical Megaverse; The Big Bang; Our Sister Universe's Wormhole; Origin of the Cosmological Constant, Spatial Asymmetry of the Universe, and its Web of Galaxies; A Baryonic Field between Universes and Particles; Megaverse Extended Wheeler-DeWitt Equation* (Blaha Reasearch, Auburn, NH, 2014).

_____, 2014b, *All the Megaverse! Starships Exploring the Endless Universes of the Cosmos Using the Baryonic Force* (Blaha Research, Auburn, NH, 2014).

_____, 2014c, *All the Megaverse! II Between Megaverse Universes: Quantum Entanglement Explained by the Megaverse Coherent Baryonic Radiation Devices – PHASERs Neutron Star Megaverse Slingshot Dynamics Spiritual and UFO Events, and the Megaverse Microscopic Entry into the Megaverse* (Blaha Research, Auburn, NH, 2014).

_____, 2015a, *PHYSICS IS LOGIC PAINTED ON THE VOID: Origin of Bare Masses and The Standard Model in Logic, U(4) Origin of the Generations, Normal and Dark Baryonic Forces, Dark Matter, Dark Energy, The Big Bang, Complex General Relativity, A Megaverse of Universe Particles* (Blaha Research, Auburn, NH, 2015).

_____, 2015b, *PHYSICS IS LOGIC Part II: The Theory of Everything, The Megaverse Theory of Everything, U(4)⊗U(4) Grand Unified Theory (GUT), Inertial Mass = Gravitational Mass, Unified Extended Standard Model and a New Complex General Relativity with Higgs Particles, Generation Group Higgs Particles* (Blaha Research, Auburn, NH, 2015).

_____, 2015c, *The Origin of Higgs ("God") Particles and the Higgs Mechanism: Physics is Logic III, Beyond Higgs – A Revamped Theory With a Local Arrow of Time, The Theory of Everything Enhanced, Why Inertial Frames are Special, Universes of the Mind* (Blaha Research, Auburn, NH, 2015).

_____, 2015d, *The Origin of the Eight Coupling Constants of The Theory of Everything: U(8) Grand Unified Theory of Everything (GUTE), S^8 Coupling Constant Symmetry, Space-Time Dependent Coupling Constants, Big Bang Vacuum Coupling Constants, Physics is Logic IV* (Blaha Research, Auburn, NH, 2015).

_____, 2016a, *New Types of Dark Matter, Big Bang Equipartition, and A New U(4) Symmetry in the Theory of Everything: Equipartition Principle for Fermions, Matter is 83.33% Dark, Penetrating the Veil of the Big Bang, Explicit QFT Quark Confinement and Charmonium, Physics is Logic V* (Blaha Research, Auburn, NH, 2016).

_____, 2016b, *The Periodic Table of the 192 Quarks and Leptons in The Theory of Everything: The U(4) Layer Group, Physics is Logic VI* (Blaha Research, Auburn, NH, 2016).

_____, 2016c, *New Boson Quantum Field Theory, Dark Matter Dynamics, Dark Matter Fermion Layer Mixing, Genesis of Higgs Particles, New Layer Higgs Masses, Higgs Coupling Constants, Non-Abelian Higgs Gauge Fields, Physics is Logic VII* (Blaha Research, Auburn, NH, 2016).

_____, 2016d, *Unification of the Strong Interactions and Gravitation: Quark Confinement Linked to Modified Short-Distance Gravity; Physics is Logic VIII* (Blaha Research, Auburn, NH, 2016).

_____, 2016e, *MoND: Unification of the Strong Interactions and Gravitation II, Quark Confinement Linked to Large-Scale Gravity, Physics is Logic IX* (Blaha Research, Auburn, NH, 2016).

_____, 2016f, *CQ Mechanics: A Unification of Quantum & Classical Mechanics, Quantum/Semi-Classical Entanglement, Quantum/Classical Path Integrals, Quantum/Classical Chaos* (Blaha Research, Auburn, NH, 2016).

_____, 2016g, *GEMS: Unified Gravity, ElectroMagnetic and Strong Interactions: Manifest Quark Confinement, A Solution for the Proton Spin Puzzle, Modified Gravity on the Galactic Scale* (Pingree Hill Publishing, Auburn, NH, 2016).

_____, 2016h, *Unification of the Seven Boson Interactions based on the Riemann-Christoffel Curvature Tensor* (Pingree Hill Publishing, Auburn, NH, 2016).

_____, 2017a, *Unification of the Eleven Boson Interactions based on 'Rotations of Interactions'* (Pingree Hill Publishing, Auburn, NH, 2017).

_____, 2017b, *The Origin of Fermions and Bosons, and Their Unification* (Pingree Hill Publishing, Auburn, NH, 2017).

_____, 2017c, *Megaverse: The Universe of Universes* (Pingree Hill Publishing, Auburn, NH, 2017).

_____, 2017d, *SuperSymmetry and the Unified SuperStandard Model* (Pingree Hill Publishing, Auburn, NH, 2017).

_____, 2017e, *From Qubits to the Unified SuperStandard Model with Embedded SuperStrings: A Derivation* (Pingree Hill Publishing, Auburn, NH, 2017).

_____, 2017f, *The Unified SuperStandard Model in Our Universe and the Megaverse: Quarks, ... ,* (Pingree Hill Publishing, Auburn, NH, 2017).

_____, 2018a, *The Unified SuperStandard Model and the Megaverse SECOND EDITION A Deeper Theory based on a New Particle Functional Space that Explicates Quantum Entanglement Spookiness (Volume 1)* (Pingree Hill Publishing, Auburn, NH, 2018).

_____, 2018b, *Cosmos Creation: The Unified SuperStandard Model, Volume 2, SECOND EDITION* (Pingree Hill Publishing, Auburn, NH, 2018).

_____, 2018c, *God Theory* (Pingree Hill Publishing, Auburn, NH, 2018).

_____, 2018d, *Immortal Eye: God Theory: Second Edition* (Pingree Hill Publishing, Auburn, NH, 2018).

_____, 2018e, *Unification of God Theory and Unified SuperStandard Model THIRD EDITION* (Pingree Hill Publishing, Auburn, NH, 2018).

_____, 2019a, *Calculation of: QED α = 1/137, and Other Coupling Constants of the Unified SuperStandard Theory* (Pingree Hill Publishing, Auburn, NH, 2019).

_____, 2019b, *Coupling Constants of the Unified SuperStandard Theory SECOND EDITION* (Pingree Hill Publishing, Auburn, NH, 2019).

_____, 2019c, *New Hybrid Quantum Big_Bang–Megaverse_Driven Universe with a Finite Big Bang and an Increasing Hubble Constant* (Pingree Hill Publishing, Auburn, NH, 2019).

_____, 2019d, *The Universe, The Electron and The Vacuum* (Pingree Hill Publishing, Auburn, NH, 2019).

_____, 2019e, *Quantum Big Bang – Quantum Vacuum Universes (Particles)* (Pingree Hill Publishing, Auburn, NH, 2019).

_____, 2019f, *The Exact QED Calculation of the Fine Structure Constant Implies ALL 4D Universes have the Same Physics/Life Prospects* (Pingree Hill Publishing, Auburn, NH, 2019).

_____, 2019g, *Unified SuperStandard Theory and the SuperUniverse Model: The Foundation of Science* (Pingree Hill Publishing, Auburn, NH, 2019).

_____, 2020a, *Quaternion Unified SuperStandard Theory (The QUeST) and Megaverse Octonion SuperStandard Theory (MOST)* (Pingree Hill Publishing, Auburn, NH, 2020).

_____, 2020b, *United Universes Quaternion Universe - Octonion Megaverse* (Pingree Hill Publishing, Auburn, NH, 2020).

_____, 2020c, *Unified SuperStandard Theories for Quaternion Universes & The Octonion Megaverse* (Pingree Hill Publishing, Auburn, NH, 2020).

_____, 2020d, *The Essence of Eternity: Quaternion & Octonion SuperStandard Theories* (Pingree Hill Publishing, Auburn, NH, 2020).

_____, 2020e, *The Essence of Eternity II* (Pingree Hill Publishing, Auburn, NH, 2020).

_____, 2020f, *A Very Conscious Universe* (Pingree Hill Publishing, Auburn, NH, 2020).

_____, 2020g, *Hypercomplex Universe* (Pingree Hill Publishing, Auburn, NH, 2020).

_____, 2020h, *Beneath the Quaternion Universe* (Pingree Hill Publishing, Auburn, NH, 2020).

_____, 2020i, *Why is the Universe Real? From Quaternion & Octonion to Real Coordinates* (Pingree Hill Publishing, Auburn, NH, 2020).

_____, 2020j, *The Origin of Universes: of Quaternion Unified SuperStandard Theory (QUeST); and of the Octonion Megaverse (UTMOST)* (Pingree Hill Publishing, Auburn, NH, 2020).

_____, 2020k, *The Seven Spaces of Creation: Octonion Cosmology* (Pingree Hill Publishing, Auburn, NH, 2020).

_____, 2020l, *From Octonion Cosmology to the Unified SuperStandard Theory of Particles* (Pingree Hill Publishing, Auburn, NH, 2020).

_____, 2021a, *Pioneering the Cosmos* (Pingree Hill Publishing, Auburn, NH, 2021).

_____, 2021b, *Pioneering the Cosmos II* (Pingree Hill Publishing, Auburn, NH, 2021).

_____, 2021c, *Beyond Octonion Cosmology* (Pingree Hill Publishing, Auburn, NH, 2021).

_____, 2021d, *Universes are Particles* (Pingree Hill Publishing, Auburn, NH, 2021).

_____, 2021e, *Octonion-like dna-based life, Universe expansion is decay, Emerging New Physics* (Pingree Hill Publishing, Auburn, NH, 2021).

_____, 2021f, *The Science of Creation New Quantum Field Theory of Spaces* (Pingree Hill Publishing, Auburn, NH, 2021).

_____, 2021g, *Quantum Space Theory With Application to Octonion Cosmology & Possibly To Fermionic Condensed Matter* (Pingree Hill Publishing, Auburn, NH, 2021).

Bogoliubov N. N., & Shirkov, D. V., Volkoff, G. M. (tr), *Introduction to the Theory of Quantized Fields* (Wiley-Interscience, New York, 1959).

Braithwaite, R. B., 1960, *Scientific Explanation* (Harper Torchbook, New York, 1960).

Carnap, R., 1956, *Meaning and Necessity* (Univ. Chicago Press, Chicago, 1956).

Carnap, R., (Ed. M. Gardner), 1995, A*n Introduction to the Philosophy of Science* (Dover Publications, New York, 1995).

Curry, H. B., 1976, *Foundations of Mathematical Logic* (Dover Publications, New York, 1976).

Davis, M., 1982, *Computability and Unsolvability* (Dover Publications, New York, 1982).

Dirac, P. A. M., 1931, *Quantum Mechanics* Third Edition (Oxford University Press, Oxford, 1947).

Eddington, A. S., 1952, *The Mathematical Theory of Relativity* (Cambridge University Press, Cambridge, U.K., 1952).

Fant, Karl M., 2005, *Logically Determined Design: Clockless System Design With NULL Convention Logic* (John Wiley and Sons, Hoboken, NJ, 2005).

Feinberg, G. and Shapiro, R., 1980, *Life Beyond Earth: The Intelligent Earthlings Guide to Life in the Universe* (William Morrow and Company, New York, 1980).

Gelfand, I. M., Fomin, S. V., Silverman, R. A. (tr), 2000, *Calculus of Variations* (Dover Publications, Mineola, NY, 2000).

Frege, G., (Ed. M. Beaney), 1997, *The Frege Reader* (Blackwell Publishing, Malden, MA, 1997).

Garson, J. W., 2006, *Modal Logic for Philosophers* (Cambridge University Press, Cambridge, 2006).

Giaquinta, M., Modica, G., Souchek, J., 1998, *Cartesian Coordinates in the Calculus of Variations* Volumes I and II (Springer-Verlag, New York, 1998).

Giaquinta, M., Hildebrandt, S., 1996, *Calculus of Variations* Volumes I and II (Springer-Verlag, New York, 1996).

Gödel, K., 1992, Tr. Meltzer, B., Introduction by R. B. Braithwaite, *On Formally Undecidable Propositions of Principia Mathematica and Related Systems* (Dover Publications, New York, 1992).

Gradshteyn, I. S. and Ryzhik, I. M., 1965, *Table of Integrals, Series, and Products* (Academic Press, New York, 1965).

Heitler, W., 1954, *The Quantum Theory of Radiation* (Claendon Press, Oxford, UK, 1954).

Huang, Kerson, 1992, *Quarks, Leptons & Gauge Fields 2nd Edition* (World Scientific Publishing Company, Singapore, 1992).

Hilbert, D. and Ackermann, W. (Tr. L. M. Hammond et al), 1950, *Principles of Mathematical Logic* (Chelsea Publishing Co., New York, 1950).

Jost, J., Li-Jost, X., 1998, *Calculus of Variations* (Cambridge University Press, New York, 1998).

Kaku, Michio, 1993, *Quantum Field Theory*, (Oxford University Press, New York, 1993).

Kirk, G. S. and Raven, J. E., 1962, *The Presocratic Philosophers* (Cambridge University Press, New York, 1962).

Kleene, S. C., 1967, *Mathematical Logic* (Dover Publications, New York, 1967).

Konyndyk, K., 1986, *Introductory Modal Logic* (University of Notre Dame Press, Notre Dame, Indiana, 1986).

Landau, L. D. and Lifshitz, E. M., 1987, *Fluid Mechanics 2nd Edition*, (Pergamon Press, Elmsford, NY, 1987).

Lavine, S., 1994, *Understanding the Infinite* (Harvard University Press, Cambridge, MA, 1994).

Loux, M. J., 2006, *Metaphysics: A Contemporary Introduction* (Routledge, New York, 2006).

Lowe, E. J., 2002, *A Survey of Metaphysics* (Oxford University Press, Oxford, 2002).

Mackey, G. W., 1963, Mathematical Foundations of Quantum Mechanics (W. A. Benjamin, New York, 1963).

Messiah, A., 1965, *Quantum Mechanics* Volume I (John Wiley & Sons, New York, 1965).

Misner, C. W., Thorne, K. S., and Wheeler, J. A., 1973, *Gravitation* (W. H. Freeman, New York, 1973).

Potter, M., 2004, *Set Theory and its Philosophy* (Oxford University Press, Oxford, 2004).

Quine, W. van O., 1962, *Mathematical Logic* (Harper Torchbooks, New York, 1962).

Rescher, N., 1967, *The Philosophy of Leibniz* (Prentice-Hall, Englewood Cliffs, NJ, 1967).

Rieffel, Eleanor and Polak, Wolfgang, 2014, *Quantum Computing* (MIT Press, Cambridge, MA, 2014).

Riesz, Frigyes and Sz.-Nagy, Béla, 1990, *Functional Analysis* (Dover Publications, New York, 1990).

Révész, G. E., 1983, *Introduction to Formal Languages* (Dover Publications, New York, 1983).

Sagan, H., 1993, *Introduction to the Calculus of Variations* (Dover Publications, Mineola, NY, 1993).

Sakurai, J. J., 1964, *Invariance Principles and Elementary Particles* (Princeton University Press, Princeton, NJ, 1964).

Smullyan, R. M., 1995, *First-Order Logic* (Dover Publications, New York, 1995).

Tarski, A., 1995, *Introduction to Logic and to the Methodology of Deductive Sciences* (Dover Publications, New York, 1995).

Tiles, M., 1989, *The Philosophy of Set Theory* (Dover Publications, New York, 1989).

Van Inwagen, P., 2009, *Metaphysics* (Westview Press, Boulder, CO, 2009).

Weinberg, S., 1972, *Gravitation and Cosmology* (John Wiley and Sons, New York, 1972).

Weinberg, S., 1995, *The Quantum Theory of Fields Volume I* (Cambridge University Press, New York, 1995).

Weyl, H., 1950, *Space, Time, Matter* (Dover, New York, 1950).

Weyl, H., (Tr. S. Pollard et al), 1987, *The Continuum* (Dover Publications, New York, 1987).

REFERENCES

INDEX

About the Author

Stephen Blaha is a well-known Physicist and Man of Letters with interests in Science, Society and civilization, the Arts, and Technology. He had an Alfred P. Sloan Foundation scholarship in college. He received his Ph.D. in Physics from Rockefeller University. He has served on the faculties of several major universities. He was also a Member of the Technical Staff at Bell Laboratories, a manager at the Boston Globe Newspaper, a Director at Wang Laboratories, and President of Blaha Software Inc. and of Janus Associates Inc. (NH).

Among other achievements he was a co-discoverer of the "r potential" for heavy quark binding developing the first (and still the only demonstrable) non-Aeolian gauge theory with an "r" potential; first suggested the existence of topological structures in superfluid He-3; first proposed Yang-Mills theories would appear in condensed matter phenomena with non-scalar order parameters; first developed a grammar-based formalism for quantum computers and applied it to elementary particle theories; first developed a new form of quantum field theory without divergences (thus solving a major 60 year old problem that enabled a unified theory of the Standard Model and Quantum Gravity without divergences to be developed); first developed a formulation of complex General Relativity based on analytic continuation from real space-time; first developed a generalized non-homogeneous Robertson-Walker metric that enabled a quantum theory of the Big Bang to be developed without singularities at t = 0; first generalized Cauchy's theorem and Gauss' theorem to complex, curved multi-dimensional spaces; received Honorable Mention in the Gravity Research Foundation Essay Competition in 1978; first developed a physically acceptable theory of faster-than-light particles; first derived a composition of extremums method in the Calculus of Variations; first quantitatively suggested that inflationary periods in the history of the universe were not needed; first proved Gödel's Theorem implies Nature must be quantum; provided a new alternative to the Higgs Mechanism, and Higgs particles, to generate masses; first showed how to resolve logical paradoxes including Gödel's Undecidability Theorem by developing Operator Logic and Quantum Operator Logic; first developed a quantitative harmonic oscillator-like model of the life cycle, and interactions, of civilizations; first showed how equations describing superorganisms also apply to civilizations. A recent book shows his theory applies successfully to the past 14 years of history and to *new* archaeological data on Andean and Mayan civilizations as well as Early Anatolian and Egyptian civilizations.

He first developed an axiomatic derivation of the form of The Standard Model from geometry – space-time properties – The Unified SuperStandard Model. It unifies all the known forces of Nature. It also has a Dark Matter sector that includes a Dark ElectroWeak sector with Dark doublets and Dark gauge interactions. It uses quantum coordinates to remove infinities that crop up in most interacting quantum field theories and additionally to remove the infinities that appear in the Big Bang and generate inflationary growth of the universe. It shows gravity has a MOND-like form without sacrificing Newton's Laws. It relates the interactions of the MOND-like sector of gravity with the r-potential of Quark Confinement. The axioms of the theory lead to the question of their origin. We suggest in the preceding edition of this book it can be attributed to an entity with God-like properties. We explore these properties in "God Theory" and show they predict that the Cosmos exists forever although individual universes (or incarnations of our universe) "come and go." Several other important results emerge from God Theory such a functionally triune God. The Unified SuperStandard Theory has many other important parts described in the Current Edition of *The Unified SuperStandard Theory* and expanded in subsequent volumes.

Blaha has had a major impact on a succession of elementary particle theories: his Ph.D. thesis (1970), and papers, showed that quantum field theory calculations to all orders in ladder approximations could not give scaling deep inelastic electron-nucleon scattering. He later showed the eigenvalue equation for the fine structure constant α in Johnson-Baker-Willey QED had a zero at $\alpha = 1$ not $1/137$ by solving the Schwinger-Dyson equations to all orders in an approximation that agreed with exact results to 4th order in α thus ending interest in this theory. In 1979 at Prof. Ken Johnson's (MIT) suggestion he calculated the proton-neutron mass difference in the MIT bag model and found the result had the wrong sign reducing interest in the bag model. These results all appear in Physical Review papers. In the 2000's he repeatedly pointed out the shortcomings of SuperString theory and showed that The Standard Model's form could be derived from space-time geometry by an extension of Lorentz transformations to faster than light transformations. This deeper space-time basis greatly increases the possibility that it is part of THE fundamental theory. Recently, Blaha showed that the Weak interactions differed significantly from the Strong, electromagnetic and gravitation interactions in important respects while these interactions had similar features, and suggested that ElectroWeak theory, which is essentially a glued union of the Weak interactions and Electromagnetism, possibly modulo unknown Higgs particle features, be replaced by a unified theory of the other interactions combined with a stand-alone Weak interaction theory. Blaha also showed that, if Charmonium calculations are taken seriously, the Strong interaction coupling constant is only a factor of five larger than the electromagnetic coupling constant, and thus Strong interaction perturbation theory would make sense and yield physically meaningful results.

In graduate school (1965-71) he wrote substantial papers in elementary particles and group theory: The Inelastic E- P Structure Functions in a Gluon Model. Phys. Lett. B40:501-502,1972; Deep-Inelastic E-P Structure Functions In A Ladder Model With Spin 1/2 Nucleons, Phys.Rev. D3:510-523,1971; Continuum Contributions To The Pion Radius, Phys. Rev. 178:2167-2169,1969; Character Analysis of U(N) and SU(N), J. Math. Phys. 10, 2156 (1969); and The Calculation of the Irreducible Characters of the Symmetric Group in Terms of the Compound Characters, (Published as Blaha's Lemma in D. E. Knuth's book: *The Art of Computer Programming Vols. 1 – 4*).

In the early 1980's Blaha was also a pioneer in the development of UNIX for financial, scientific and Internet applications: benchmarked UNIX versions showing that block size was critical for UNIX performance, developing financial modeling software, starting database benchmarking comparison studies, developing Internet-like UNIX networking (1982) and developing a hybrid shell programming technique (1982) that was a precursor to the PERL programming language. He was also the manager of the AT&T ten-year future products development database. His work helped lead to commercial UNIX on computers such as Sun Micros, IBM AIX minis, and Apple computers.

In the 1980's he pioneered the development of PC Desktop Publishing on laser printers and was nominated for three "Awards for Technical Excellence" in 1987 by PC Magazine for PC software products that he designed and developed.

Recently he has developed a theory of Megaverses – actual universes of which our universe is one – with quantum particle-like properties based on the Wheeler-DeWitt equation of Quantum Gravity. He has developed a theory of a baryonic force, which had been conjectured many years ago, and estimated the strength of the force based on discrepancies in measurements of the gravitational constant G. This force, operative in D-dimensional space, can be used to escape from our universe in "uniships" which are the equivalent of the faster-than-light starships proposed in the author's earlier books. Thus travel to other universes, as well as to other stars is possible.

Blaha also considered the complexified Wheeler-DeWitt equation and showed that its limitation to real-valued coordinates and metrics generated a Cosmological Constant in the Einstein equations.

The author has also recently written a series of books on the serious problems of the United States and their solution as well as a book on the decline of Mankind that will follow from current social and genetic trends in Mankind.

In the past twenty years Dr. Blaha has written over 80 books on a wide range of topics. Some recent major works are: *From Asynchronous Logic to The Standard Model to Superflight to the Stars, All the Universe!*, *SuperCivilizations: Civilizations as Superorganisms*, *America's Future: an Islamic Surge, ISIS, al Qaeda, World Epidemics, Ukraine, Russia-China Pact, US Leadership Crisis*, *The Rises and Falls of Man – Destiny – 3000 AD: New Support for a Superorganism MACRO-THEORY of CIVILIZATIONS From CURRENT WORLD TRENDS and NEW Peruvian, Pre-Mayan, Mayan, Anatolian, and Early Egyptian Data, with a Projection to 3000 AD*, and *Mankind in Decline: Genetic Disasters, Human-Animal Hybrids, Overpopulation, Pollution, Global Warming, Food and Water Shortages, Desertification, Poverty, Rising Violence, Genocide, Epidemics, Wars, Leadership Failure.*

He has taught approximately 4,000 students in undergraduate, graduate, and postgraduate corporate education courses primarily in major universities, and large companies and government agencies.

Recently he developed a quantum theory, The Unified SuperStandard Theory (UST), which describes elementary particles in detail without the difficulties of conventional quantum field theory. He found that the internal symmetries of this theory could be exactly derived from an octonion theory called QUeST. He further found that another octonion theory (UTMOST) describes the Megaverse. It can hold QUeST universes such as our own universe. It has an internal symmetry structure which is a superset of the QUeST internal symmetries. This book provides a complete derivation from the Quantum Field Theory of Spaces.